IUTAM Bookseries

Volume 33

The IUTAM Bookseries publishes the refereed proceedings of symposia organised by the International Union of Theoretical and Applied Mechanics (IUTAM).

Every two years the IUTAM General Assembly decides on the list of IUTAM Symposia. The Assembly calls upon the advice of the Symposia panels. Proposals for Symposia are made through the Assembly members, the Adhering Organizations, and the Affiliated Organizations, and are submitted online when a call is launched on the IUTAM website.

The IUTAM Symposia are reserved to invited participants. Those wishing to participate in an IUTAM Symposium are therefore advised to contact the Chairman of the Scientific Committee in due time in advance of the meeting. From 1996 to 2010, Kluwer Academic Publishers, now Springer, was the preferred publisher of the refereed proceedings of the IUTAM Symposia. Proceedings have also been published as special issues of appropriate journals. From 2018, this bookseries is again recommended by IUTAM for publication of Symposia proceedings.

Indexed in Ei Compendex and Scopus.

More information about this series at http://www.springer.com/series/7695

Evtim Zahariev • Javier Cuadrado
Editors

IUTAM Symposium on Intelligent Multibody Systems – Dynamics, Control, Simulation

Springer

Editors
Evtim Zahariev
Institute of Mechanics
Bulgarian Academy of Sciences
Sofia, Bulgaria

Javier Cuadrado
Laboratory of Mechanical Engineering
University of La Coruña
Ferrol, Spain

ISSN 1875-3507 ISSN 1875-3493 (electronic)
IUTAM Bookseries
ISBN 978-3-030-00526-9 ISBN 978-3-030-00527-6 (eBook)
https://doi.org/10.1007/978-3-030-00527-6

Library of Congress Control Number: 2018962961

This Springer imprint is published by the registered company Springer Nature Switzerland AG
The registered company address is: Gewerbestrasse 11, 6330 Cham, Switzerland

Preface

Multibody dynamics as independent branch of mechanics was agreed to and established in 1977 at an IUTAM Symposium held in Munich. Recently, the multibody system methodology experienced huge progress including real-time simulation, contact and impact problems, extension to electronics and mechatronics, dynamic strength analysis, optimization of design and control devices, integration codes in particular for differential algebraic systems, challenging applications in biomechanics, robotics and vehicle dynamics, manufacturing science and molecular dynamics, etc.

Examples of such mechatronic (intelligent) multibody systems are autonomous vehicles, space structures, structures exposed to external and seismic excitations, large flexible structures and wind generators, robots and bio-robots, manufacturing systems, etc.

Solution of these tasks requires up-to-date methods for dynamics analysis and simulation, novel methods for numerical solution of ODE and DAE, real-time simulation, and passive, semi-passive, and active control algorithms. The topics of the colloquium are:

- Novel methods in multibody system dynamics;
- Real-time dynamics;
- Dynamic models of passive and active mechatronic devices;
- Vehicle dynamics and control; -Structural dynamics;
- Deflection and vibration suppression;
- Numerical integration of ODE and DAE for large scale and stiff multibody systems;
- Model reduction of large-scale flexible systems.

The symposium gathered respected scientist in the field of multibody systems dynamics that extended their knowledge and experience to complicated problems and applications. Twenty-five participants from 13 countries (Bulgaria, Germany, Belgium, the USA, Mexico, Macao, China, Spain, Iran, the UK, India, Croatia, Austria) attended the event.

Sofia, Bulgaria — Evtim Venets Zahariev
Ferrol, Spain — Javier Cuadrado

Scientific Committee

Chair
Prof. Evtim Zahariev
Institute of Mechanics
Bulgarian Academy of Sciences
Sofia, Bulgaria

Members
Prof. Jorge Ambrosio
IDMEC/IST
Instituto Superior Tecnico
Lisboa, Portugal

Prof. Javier Cuadrado
Laboratory of Mechanical Engineering
University of La Coruña
Ferrol, Spain

Prof. Olivier Brüls
Department of Aerospace and Mechanical Engineering (LTAS)
University of Liège
Liège, Belgium

Prof. Andreas Müller
University of Linz
Johannes Kepler University
Linz, Austria

Prof. Ahmed Shabana
Department of Mechanical Engineering
University of Illinois at Chicago
Chicago, Illinois
USA

Prof. John McPhee
Systems Design Engineering Department
University of Waterloo
Waterloo, Ontario
Canada

Prof. Georgi Todorov
Technical University Sofia
Sofia, Bulgaria

IUTAM Representative on Scientific Committee

Prof. Peter Eberhard
University of Stuttgart
Germany

Organizing Committee

Prof. Evtim Zahariev
Institute of Mechanics
Bulgarian Academy of Sciences
Sofia, Bulgaria

Members
Prof. Kamen Delchev
Institute of Mechanics
Bulgarian Academy of Sciences
Sofia, Bulgaria

Assist. Prof. Tsvetelina Yorgova
Institute of Mechanics
Bulgarian Academy of Sciences
Sofia, Bulgaria

Assist. Prof. Stefan Karastanev
Institute of Mechanics
Bulgarian Academy of Sciences
Sofia, Bulgaria

Ph.D. Ilia Savov
Technical University Sofia
Sofia, Bulgaria

Student Boris Nentchovski
Technical University Sofia
Sofia, Bulgaria

Reviewers

- Prof. Javier Cuadrado, Laboratory of Mechanical Engineering, University of La Coruña, Ferrol, Spain
- Prof. John McPhee, Systems Design Engineering Department, University of Waterloo, Waterloo, Ontario, Canada
- Prof. Ahmed Shabana, Department of Mechanical Engineering, University of Illinois at Chicago, Chicago, USA
- Prof. Jorge Ambrosio, Instituto Superior Tecnico, Lisboa, Portugal
- Prof. Juan Carlos Garcia Orden, Universidad Politecnica de Madrid, Departmento de Mecanica de Medios Continuos y Teoria de Estructuras, Madrid, Spain
- Prof. Josep Font Llagunes, Universitat Politècnica de Catalunya, Escola Tècnica Superior d'Enginyeria Industrial de Barcelona, Spain
- Prof. Subir Saha, Department of Mechanical Engineering, IIT, Delhi, New Delhi, India
- Prof. Andreas Müller, Johannes Kepler University of Linz, Linz, Austria
- Prof. Luciano Menegaldo, Biomedical Engineering Program, Federal University of Rio de Janeiro, Rio de Janeiro, Brazil
- Prof. Peter Eberhard, Institute of Engineering and Computational Mechanics, University of Stuttgart, Germany
- Prof. Olivier Brüls, Department of Aerospace and Mechanical Engineering, University of Liège, Liège, Belgium
- Prof. Radu Serban, Department of Mechanical Engineering, University of Wisconsin, Madison, USA
- Prof. Petko Kiriazov, Institute of Mechanics, Bulgarian Academy of Sciences, Bulgaria
- Prof. Evtim Zahariev, Institute of Mechanics, Bulgarian Academy of Sciences, Sofia, Bulgaria

Contents

Contributors

Jorge Ambrósio IDMEC, Instituto Superior Técnico, University of Lisbon, Lisbon, Portugal

David T. Branson Advanced Manufacturing Technology Research Group, Faculty of Engineering, University of Nottingham, Nottingham, UK

Olivier Brüls Université de Liège, Liège, Belgium

Javier Cuadrado Laboratory of Mechanical Engineering, University of La Coruña, Ferrol, Spain

Petia Dineva Bulgarian Academy of Sciences, Institute of Mechanics, Sofia, Bulgaria

Quentin Docquier Université Catholique de Louvain, Louvain-la-Neuve, Belgium

Hai-En Du University of Macau, Macau SAR, People's Republic of China

Saeed Ebrahimi Mechanical Engineering Department, Yazd University, Yazd, Iran

Guo-Kang Er University of Macau, Macau SAR, People's Republic of China

Paul Fisette Université Catholique de Louvain, Louvain-la-Neuve, Belgium

Hubert Gattringer Institute of Robotics, Johannes Kepler University Linz, Linz, Austria

Isuru S. Godage School of Computing, DePaul University, Chicago, IL, USA

Hossein Habibi School of Science, Engineering and Design, Teesside University, Middlesbrough, UK

Vai Pan Iu University of Macau, Macau SAR, People's Republic of China

Konstantin Kamberov Faculty of Industrial Technology, Technical University Sofia, Sofia, Bulgaria

Rongjie Kang Key Laboratory of Mechanism Theory and Equipment Design of Ministry of Education, School of Mechanical Engineering, Tianjin University, Tianjin, China

Urbano Lugris Laboratory of Mechanical Engineering, University of La Coruña, Ferrol, Spain

Arman Mardani Mechanical Engineering Department, Yazd University, Yazd, Iran

Florian Michaud Laboratory of Mechanical Engineering, University of La Coruña, Ferrol, Spain

Francisco Mouzo Laboratory of Mechanical Engineering, University of La Coruña, Ferrol, Spain

Andreas Müller Institute of Robotics, Johannes Kepler University Linz, Linz, Austria

Florian Pucher Institute of Robotics, Johannes Kepler University Linz, Linz, Austria

Alexander Reiter Institute of Robotics, Johannes Kepler University Linz, Linz, Austria

Subir Kumar Saha Indian Institute of Technology Delhi, New Delhi, India

Iliya Savov Faculty of Industrial Technology, Technical University Sofia, Sofia, Bulgaria

Sasanka Sekhar Sinha Indian Institute of Technology Delhi, New Delhi, India

Yavor Sofronov Faculty of Industrial Technology, Technical University Sofia, Sofia, Bulgaria

Georgi Todorov Faculty of Industrial Technology, Technical University Sofia, Sofia, Bulgaria

Ian D. Walker Department of Electrical & Computer Engineering, Clemson University, Clemson, SC, USA

Kun Wang University of Macau, Macau SAR, People's Republic of China

Evtim Zahariev Institute of Mechanics, Bulgarian Academy of Sciences, Sofia, Bulgaria

List of Sponsors

- Institute of Mechanics, Bulgarian Academy of Sciences
- Bulgarian Academy of Sciences
- IFToMM: International Federation for the Promotion of Mechanism and Machine Science
- Gesellschaft für Angewandte Mathematik und Mechanik
- Faculty of Industrial Technology, Technical University Sofia

Chapter 1
Selected Challenges in Realistic Multibody Modeling of Machines and Vehicles

Jorge Ambrósio

Abstract Multibody modelling involves taking fundamental decisions during the multibody model construction that not only condition its validity for the particular application foreseen but also have fundamental implications on the suitability of the numerical methods used on its analysis. The decision on allowing a particular flexible body to exhibit linear or nonlinear deformations or even to consider it part of the multibody system or as a structural component with which the system interacts is a crucial part of the modeling process. The description of the kinematic relations between moving components can be represented by perfect kinematic constraints or by contact pairs, as when local effects in the joints, generally associated with deviations from nominal conditions or to functional features, must be considered. The interaction of the multibody model with the 'environment' may require a substantial modelling effort that involves decisions on geometric description of features, on contact mechanics or even on numerical methodologies to handle co-simulation of systems with equilibrium equations that are solved with different numerical methods. Several areas are transversal to all the challenges identified and discussed here. The suitability of the numerical time integrators not only to handle the multibody model assumptions but also their interaction with other systems are of fundamental importance in the correct solution of the system dynamics. Descriptive and differential geometry also plays a very important role not only in the description of the relative kinematics of the systems but also in the modelling of the interactions.

Keywords Geometry · Look-up-table · Clearance joints · Contact mechanics · Numerical integration

---

J. Ambrósio (✉)
IDMEC, Instituto Superior Técnico, University of Lisbon, Lisbon, Portugal
e-mail: Jorge.Ambrosio@tecnico.ulisboa.pt

E. Zahariev, J. Cuadrado (eds.), *IUTAM Symposium on Intelligent Multibody Systems – Dynamics, Control, Simulation*, IUTAM Bookseries 33,
https://doi.org/10.1007/978-3-030-00527-6_1

1.1 Introduction

The early developments of multibody dynamics focused the dynamics of spacecrafts, modelled as rigid bodies [1] or the biomechanics of the human body for space applications [2, 3]. Among others, the work of Wittenburg addressed the dynamics of coupled bodies [4] and originated the first software tool based on multibody dynamics methods for simulating the dynamics injury biomechanics in car accidents, the software tool MESA VERDE (MEchanism, SAtellite, VEhicle, Robot Dynamics Equations) [5]. The developments in multibody dynamics continued being reported in important meetings, started by that organized by Magnus [6], in which the term Multibody Dynamics was coined, and followed by the NATO-ASI on computer-aided analysis and optimization of multibody systems organized by Haug in 1983 [7], the IUTAM symposium on dynamics of multibody systems organized by Bianchi and Schiehlen in 1985 [8] and the NATO-ASI on multibody dynamics organized by Pereira and Ambrosio in 1993 [9]. In a review of the State-of-Art of multibody dynamics, in 1997, Schiehlen [10] identified most of the formalisms used today, as being developed already. Textbooks published by Wittenburg [11], Nikravesh [12], Kane et al. [13], Haug [14] or Garcia de Jalon and Bayo [15], describe such formalisms in a comprehensive manner. The inspiring applications of multibody dynamics, in particular to vehicle dynamics, general machinery and aerospace, and challenges for new developments, which included data models for CAD, parameter optimization for surrogate models, optimal design of multibody systems, strength analysis, i.e., flexible multibody dynamics, contact/impact problems, mechatronics or multiphysics problems, in which the interaction with fluids posed important challenges, were identified by Schiehlen as challenges for the community to be addressed [10]. Also the numerical issues associated with time integration codes or with real-time simulation were already deemed as challenges for the development of multibody systems at the time [10], and still remain important challenges today.

A major challenge that soon emerged in multibody dynamics involved the deformations of the components of the system. In most of the earlier work dealing with the elastodynamics of mechanical systems the deformations of the system components, assumed elastic and small, is superimposed to their large rigid body motion, as observed in different reviews, Erdman and Sandor [16], Lowen and Chassapis [17] and Thompson and Sung [18]. Using reference frames fixed to planar flexible bodies, Song and Haug [19] suggest a finite element based methodology, which yields coupled gross rigid body motion and small elastic deformations, which is further developed and generalized by Shabana and Wehage [20, 21] that use substructuring and the mode component synthesis to reduce the number of generalized coordinates required to represent the flexible components. The community studying space dynamics was naturally involved in dealing with the dynamics of flexible bodies undergoing large rigid body motion. The problem that attracted their initial attention was the stabilization of spinning spacecrafts with flexible appendages [22]. The need to characterize dynamically and control such systems, in particular, motivated valuable investigations on flexible multibody dynamic [23, 24]. In the framework of the spinning spacecrafts modeling, Kane, Ryan and Banerjee [25] showed that though most of the flexible multibody methods, at the time, could capture the inertia coupling between the elastodynamics of the system components and their large motion but they would still produce incorrect results because they neglected the dynamic stiffening effects. This comment and the research work, in the years that followed, addressing the nature and the solutions of such problem [26] also fostered the need for more systematic approaches to deal with nonlinear deformations in flexible multibody dynamics. Although the floating reference frame methods used in flexible multibody dynamics have the ability to lower the geometric nonlinearities of the flexible bodies they do not eliminate them because the moderate rotation assumption about the floating reference frame is still required [27]. The work of researchers in the finite element community, such as that by Belytschko and Hsieh [28], Simo and Vu-Quoc [29] or Bathe and Bolourchi [30] among others, addressing the same type of problems dealt with a similar problem. Recognizing the problem posed and using some of the approaches well in line with those of the finite element community Cardona and Geradin proposed formulations for the nonlinear flexible bodies using either a geometrically exact model [31] or through substructuring [32]. Defining it as an absolute nodal coordinate formulation, Shabana [33] used finite rotations nodal coordinates enabling the capture of the geometric nonlinear deformations. Another approach taken by Ambrósio and Nikravesh [34] to model geometrically nonlinear flexible bodies was to relax the need for the structures to exhibit small moderate rotations about the floating frame by using an incremental finite element approach within the flexible body description. The approach is further extended to handle material nonlinearities of flexible multibody systems also [35] which is applied to different types of nonlinear flexible body systems including vehicle crash dynamics [36]. In a review of the State-of-Art in flexible multibody dynamics Shabana [37] identified most of the formalisms to describe flexible bodies used today and presented views on, what were at the time, current and future research directions. The topics on nonlinear deformations of flexible bodies, on the stiff nature of the flexible body equations of motion due to the coexistence of high and low frequency components in the system response, on the use of flexible multibody dynamics in the context of computer graphics or the raise of importance of these tools in the field of biomechanics were some of research directions identified. A good part of the current approaches to flexible multibody dynamics is presented in textbooks such as those by Geradin and Cardona [38], Shabana [20], Bauchau [39] or Bremer [40].

In both rigid and flexible multibody systems the correct representation of the kinematic or kinetic relations between different bodies is of fundamental importance for the identification of the correct system dynamic response. The traditional multibody formalisms handle such relative motion restrictions as kinematic constraints [11–15, 20, 38–40]. The complexity of such kinematic restrictions, or kinematic joints, may be relatively small, as for the most common spherical, revolute or translations joints, or extremely high such as in the case of spatial cam joints, as those studied by Gonzales-Palacios and Angeles [41, 42] or path-follower constraints used in roller coaster applications, as those proposed by Pombo and Ambrosio [43, 44] or by Tandl and Kecskemethy [45, 46]. In a real mechanical joint a gap is always present, being necessary not only to allow the relative motion between the connected bodies but also to permit the components assemblage. No matter how small the clearance is, it can lead to vibration and fatigue phenomena, wear, and lack of precision or even random overall behavior [47]. However, clearance joints also allow for the use of bushings, fundamental to control the vibration in some systems and to allow small misalignments between connected bodies that, otherwise, would not exhibit the correct mobility [48, 49]. A general formulation for joints with clearance and/or bushing elements, eventually having joint motion stops, was proposed by Ambrosio and Pombo [50]. Using an approach based on non-smooth mechanics Akhadkar et al. also addressed the modelling of clearance joints with efficiency [51]. In the framework of rigid multibody systems a large library of kinematic joints, with or without clearances or bushings, has been developed using the coordinates of choice of the developers. The introduction of flexible bodies in the system involves using new sets of generalized coordinates to describe the flexibility. Even if the finite element method is used to describe the system flexibility, depending on the formulation in flexible multibody dynamics any kinematic joint available in a multibody code must be implemented with the new generalized coordinates if it is to be used in the model construction [52]. If the joint axis, itself, is flexible there is no way around it and it has to be formulated using the set of generalized coordinates involved in the formulation [53, 54]. However, if there are no local deformation effects in the joint itself, except for those associated with bushings, it is advantageous that all library of joints, kinematically perfect or represented by clearance/bushings, can be used without requiring them to be formulated with the new generalized coordinates. The use of virtual bodies, i.e., massless rigid bodies fixed to in the locations of the flexible bodies in which kinematic joints need to be connected, proposed by Bae et al. [55] and further developed by Gonçalves and Ambrosio [56], solves the problem with computational efficiency.

Although some challenges remain in terms of formulation of rigid and flexible multibody systems and on the kinematic or contact description of the joints, the identification of the major challenges has been driven by applications associated to vehicle dynamics, both road and railway, biomechanical systems, mechatronics, spacecrafts or high-precision machinery. Transversal to all areas in which challenges in multibody dynamics are identified is the computational geometry for description of curves and surfaces, required in the formulation of all kinds of interactions involving rigid or flexible bodies. Inheriting part of the framework of their formalisms from the computational geometry, the efficient formulation of complex kinematic joints is highly dependent on the efficient use of the geometric information, being this a challenge to be addressed. Contact mechanics, which always played a very important role in the dynamics of multibody systems, bases its formulations on the correct solution of the contact detection problem, which basically involves comparing geometries. In both kinematics and contact the requirements for the geometric formulation of the curves and surfaces are highly dependent on the formulations used in their implementation and on the numerical methods used for the solution. In the framework of the formulation of clearance/bushing joints all issues associated with geometry and contact mechanics are inherited and play important roles in the accurate and efficient descriptions. The use of online calculations or of Look-up-Tables (LuT), associated to the geometry of the interacting bodies in the framework of kinematics or contact, may also be determinant for the computational efficiency of a multibody program. These are of particular interest in application in which the surfaces or trajectories are very long over a non-regular geometry, being the information contained in the LuT associated with position, normal and tangent vectors and their derivatives as function of the curve or surface defining parameters, in kinematics or in contact detection, or with reaction forces and moments, in the solution of the contact law in contact mechanics. The implication of the different strategies in dealing with the geometric aspects in the context of multibody dynamics is directly observed in the behavior of the time integrator schemes used for the dynamic analysis. Therefore, the computational geometry challenges associated to kinematic formulation of spatial joints of the path-following type and contact mechanics, are addressed here. In the process, the relation between the requirements for the description of geometric features and the numerical methods used are handled with particular attention.

1.2 Geometry in Multibody Dynamics

Multibody dynamics involves the interaction between bodies, regardless of these being rigid or flexible, described by contact pairs or by restrictions of their relative motion. In order to have a computational efficient description of the body-to-body interaction the description of the geometry of the interacting curves or surfaces has to be computationally efficient, to be smooth, to have the correct degree of continuity and to allow establishing all quantities required to the proper description of the phenomena. The interaction between two bodies can be described by a kinematic constraint, such as those in cams [41], point follower mechanisms [42] or in roller coasters [43, 46]. In this case not only the description of the curve, or surface, geometry, including its normal and tangent vectors, is required but also the derivatives of all these quantities with respect to the parameters used for their description. When the interaction is described as a contact problem, such as in contact/impact between different bodies [57–60], description of biological articular surfaces [61–63] or in the formulation of clearance joints [47–50], the need to higher derivatives of the normal and tangent vectors to the surfaces or curves may not exist. In turn, due to the non-smoothness of the problem, the demand for computational efficiency is extremely high requiring that information on the geometry is properly stored and interpolated or solved online.

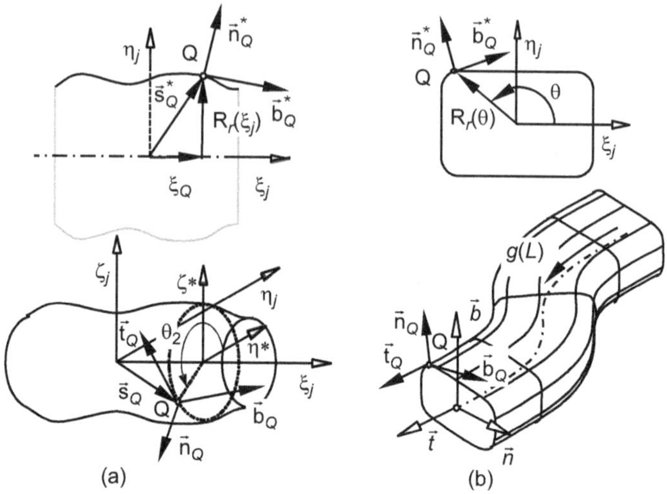

Fig. 1.1 Surfaces resulting from sweeping planar curves about axis or along spatial curves: (**a**) Surface of revolution; (**b**) Extrusion surface

1.2.1 Geometry of Curves and Surfaces

In a wide number of multibody systems bodies surface geometries that result from the sweeping of a cross section around an axis, known as surfaces of revolution, or along a curve, known as extrusions or sweep surfaces, play important roles [64]. Figure 1.1 presents general forms of bodies geometries resulting from sweeping planar curves.

The surface of revolution is obtained by rotating a plane line about an axis of revolution, as in Fig. 1.1a. The coordinates of any point in the surface are expressed in terms of the parameters defining the planar line, $R(\xi_j)$, and the sweep angle, is θ_2. In the $(\xi\eta)_j$ plane the position of point Q is a function of a single parameter, the coordinate ξ_Q, as,

$$\mathbf{s}_Q^* = \begin{Bmatrix} \xi_Q \\ R(\xi_Q) \\ 0 \end{Bmatrix} \tag{1.1}$$

The normal and tangent vectors to point Q in the line, in Fig. 1.1a, are also functions of the parameter defining the line as,

$$\mathbf{n}_Q^* = \begin{Bmatrix} n_{\xi_Q}^*(\xi_Q) \\ n_{\eta_Q}^*(\xi_Q) \\ 0 \end{Bmatrix} \; ; \; \mathbf{b}_Q^* = \begin{Bmatrix} b_{\xi_Q}^*(\xi_Q) \\ b_{\eta_Q}^*(\xi_Q) \\ 0 \end{Bmatrix} \tag{1.2}$$

The surface of revolution is obtained by rotating the line around the axis of revolution ξ_j. This operation is a transformation of coordinates in which any point on the line, given by Eq. (1.1), and the normal and tangent vectors, described in Eq. (1.2), are rotated as,

$$\begin{aligned} \mathbf{s}'_Q &= \mathbf{A}_{\theta_2}\mathbf{s}^*_Q \\ \mathbf{n}'_Q &= \mathbf{A}_{\theta_2}\mathbf{n}^*_Q \\ \mathbf{b}'_Q &= \mathbf{A}_{\theta_2}\mathbf{b}^*_Q \end{aligned} \tag{1.3}$$

in which the rotation matrix $\mathbf{A}_{\theta_2}$ is defined as,

$$\mathbf{A}_{\theta_2} = \begin{bmatrix} 1 & 0 & 0 \\ 0 & \cos\theta_2 & -\sin\theta_2 \\ 0 & \sin\theta_2 & \cos\theta_2 \end{bmatrix} \tag{1.4}$$

The point Q position on the surface of revolution and the vectors associated to the surface, given by Eq. (1.3) are expressed in the body j coordinate system. The tangent to the surface on point Q, $\mathbf{t}'_Q$, is perpendicular to both $\mathbf{n}'_Q$ and $\mathbf{b}'_Q$, being calculated by using the cross product between these vectors.

An extrusion, or sweep, surface is obtained by sweeping a planar line, and eventually rotating it, along a spatial line. Take the line shown in Fig. 1.1b, for which the position of point Q, in the $(\xi\eta)_j$ plane, is a function of a single parameter, θ, being written as,

$$\mathbf{s}^*_Q = R(\theta)\begin{Bmatrix} \cos\theta \\ \sin\theta \\ 0 \end{Bmatrix} \tag{1.5}$$

The normal and tangent vectors to point Q in the line, in Fig. 1.1b, also functions of the parameter defining the line, are written in a generic form as in Eq. (1.2).

Let the curve along which the planar curve is swept, shown in Fig. 1.2b, be defined in body j coordinates as function of an arc-length parameter L. Let a moving coordinate frame, defined by vectors $\mathbf{n}$, $\mathbf{b}$ and $\mathbf{t}$. The position of a point Q in the surface and the corresponding normal and tangent vectors are now written as

$$\begin{aligned} \mathbf{s}'_Q &= \mathbf{g}'(L) + \mathbf{A}_r\mathbf{s}^*_Q \\ \mathbf{n}'_Q &= \mathbf{A}_r\mathbf{n}^*_Q \\ \mathbf{b}'_Q &= \mathbf{A}_r\mathbf{b}^*_Q \end{aligned} \tag{1.6}$$

in which the rotation matrix $\mathbf{A}_r$ is defined as,

$$\mathbf{A}_r = \left[\mathbf{n}'(L)\ \ \mathbf{b}'(L)\ \ \mathbf{t}'(L)\right] \tag{1.7}$$

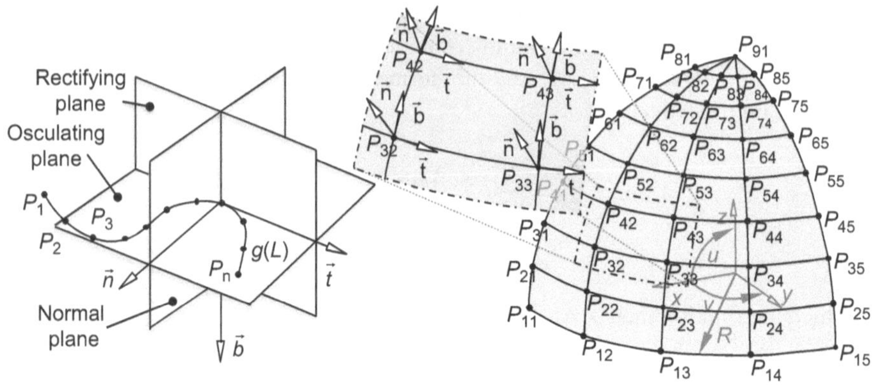

Fig. 1.2 Parametric representations of curves and surfaces: (**a**) Spatial curve; (**b**) Freeform surface

where all vectors are expressed in the body *j* coordinate system, i.e., in the coordinate system of the body to which the spatial line is attached.

A general form of representing curves and surfaces implies a parametric description in which a single parameter is enough for a curve, for instance parameter *L* in Fig. 1.2a, and two parameters are necessary for a surface, as parameters *u* and *v* for the surface in Fig. 1.2b. Then, the position of the points that belong to the curve, or the surface, and the normal and tangent vectors are explicitly obtained by mappings those parameters into a 3D-space.

The spatial curve be described using an nth order spline segments, interpolating a set of control points, defined as

$$\mathbf{g}(u) = \begin{Bmatrix} x(u) \\ y(u) \\ z(u) \end{Bmatrix} = a_0 + a_1 u + a_2 u^2 + a_3 u^3 + \cdots + a_n u^n \tag{1.8}$$

where $\mathbf{g}(u)$ is the vector locating a point on the curve, u is the local parametric variable and a_i are unknown algebraic coefficients that must be calculated using points with known coordinates. The curve moving frame, described by the Frenet frame in this work, the starts with the identification of the osculating plane, at a given point P on a curve, which is the plane of closest contact to the curve in the neighborhood of P [65]. The tangent vector $\mathbf{t}$ and the principal normal vector $\mathbf{n}$ are defined in the osculating plane. The binormal vector $\mathbf{b}$ is defined as being normal to the other two vectors, as shown in Fig. 1.2a. These vectors are defined in the intersection of the normal, rectifying and osculating planes, point P, and can be written as [43]

$$\mathbf{t} = \frac{\mathbf{g}^u}{\|\mathbf{g}^u\|} \;;\; \mathbf{n} = \frac{\mathbf{k}}{\|\mathbf{k}\|} \;;\; \mathbf{b} = \tilde{\mathbf{t}}\mathbf{n} \tag{1.9}$$

where $\tilde{\mathbf{t}}\mathbf{n}$ means a cross product and the auxiliary vector $\mathbf{k}$ is given by

$$\mathbf{k} = \mathbf{g}^{uu} - \frac{{\mathbf{g}^{uu}}^T \mathbf{g}^u}{\|\mathbf{g}^u\|^2} \mathbf{g}^u \tag{1.10}$$

in which $\mathbf{g}^u$ and $\mathbf{g}^{uu}$ are, respectively, the first and second derivatives of the parametric curve $\mathbf{g}(u)$ with respect to the parametric variable u.

The parametric surface, shown in Fig. 1.2b, can be expressed by parameters u and v that map into a 3D space. Without lack of generality, take the example of a spherical cap, which has the parametric representation given by

$$s(u, v) = \begin{cases} x = R\sqrt{\cos^2 u \cos^2 v} \\ y = R\sqrt{\cos^2 u \sin^2 v} \\ z = R\sqrt{\sin^2 u} \end{cases} \quad \begin{matrix} 0 \le u \le \frac{\pi}{2} & [\text{rad}] \\ 0 \le v < \frac{\pi}{2} & [\text{rad}] \end{matrix} \tag{1.11}$$

For each point in the surface, tangent vectors to the u and v directions, referred to as tangent and bitangent vectors respectively,

$$\mathbf{t}^u \equiv \mathbf{t}^u(u, v) = \frac{\partial \mathbf{s}(u, v)}{\partial u}, \quad \mathbf{t}^v \equiv \mathbf{t}^v(u, v) = \frac{\partial \mathbf{s}(u, v)}{\partial v} \tag{1.12}$$

$$\mathbf{n} \equiv \mathbf{n}(u, v) = \frac{\tilde{\mathbf{t}}^u \mathbf{t}^v}{\left\|\tilde{\mathbf{t}}^u \mathbf{t}^v\right\|} \tag{1.13}$$

where the tilde (~) over a vector indicates that its components are used for the skew-symmetric matrix used for a cross product [12]. To obtain the orthonormal referential associated to each contact point, the tangent vector $\mathbf{t}$ and the binormal vector $\mathbf{b}$ are

$$\mathbf{t} = \mathbf{t}^u / \left\|\mathbf{t}^u\right\|, \quad \mathbf{b} = \tilde{\mathbf{n}}\mathbf{t} \tag{1.14}$$

Note that in the calculation of the normal vector it assumed that there is no parametrization degeneration and, therefore, $\mathbf{t}^u$ and $\mathbf{t}^v$ are not parallel.

The parametrizations defined by Eqs. (1.1) and (1.4), for the spatial curve and surface respectively, do not ensure that these have constant velocity, i.e., that equal increases of a parameter in two different parts of the geometric shape lead to curve segments with different arc-length. Before the geometries can be used the parameters u, defining the curve, and u and v, defining the surface, need to be replaced by curve arc-length parameters L_u, for the curve, and L_u and L_v, for the surface, with respect to which the interpolating polynomial has a constant velocity. For instance, the parametric variable u^P of a curve, corresponding to a point P, located on the kth polynomial segment is associated to a curve length L_k^P measured from the kth segment origin, such that [43]

$$\int_{0}^{u^P} \sqrt{g_k^{u^T} g_k^u} du - L_k^P = 0 \tag{1.15}$$

In terms of its computer implementation, the non-linear Eq. (1.15) is solved in the program pre-processor, using Newton-Raphson method [12]. For the freeform surface, a similar procedure must be implemented for its re-parametrization.

For contact applications the evaluation of the positions of the points in the geometry and the corresponding normal and tangential vectors is often enough. However, when used to define kinematic constraints the derivatives of the normal and tangential vectors are also required. Therefore, the interpolation of curves and surfaces have minimum requirements in terms of the order of the interpolating polynomials that allow calculating all derivatives used in the formulation of the kinematic constraints [44].

1.2.2 Lookup Tables for Computational Efficiency

In a general computer implementation, it may be too costly to evaluate online the point positions, vectors and vector derivatives required by a particular problem. Take the case of a general contact problem of a multibody system in which there are multiple contact pairs between different surfaces, as in the case of rail-wheel contact for railway vehicles, or the case of a multibody system in which kinematic constraints are defined between curves and bodies, as in a roller coaster application [43]. Instead performing all calculations of the spatial geometry online it is computationally more efficient to use lookup tables (LuT) in which all quantities are pre-calculated as function of the curves and surfaces parameters and, at most, interpolated linearly [66].

After selecting any of the parametric descriptions of the spatial curve, the length parameter step ΔL_u adopted for the database construction is chosen, such a way that the any linear interpolation leads to an error, with respect to the exact value of any of the table contents, below a specified threshold [43]. The geometric parameters are organized in 37 columns as function of the length parameter L_u of the curve, as shown in Fig. 1.3.

Assume any freeform surface interpolation scheme for the surface and a re-parameterization with the arc-length parameter steps ΔL_u and ΔL_v adopted for the database construction. Also assume that the length parameter steps are selected such a way that the bi-linear interpolation leads to interpolation errors smaller then a specified threshold. The geometric parameters are organized in $1+\Delta L_v \times 12$ columns as function of the length parameter L_u of the curve, as shown in Fig. 1.4.

There are two clear problems with the LuT organized in the form described in Fig. 1.3 and in Fig. 1.4: The sequential access to the surface LuT is computationally inefficient when progressing along the arc-length parameter L_v, and for long curves,

L_u	For constraints and contact				For kinematic constraints only							
	$\mathbf{g}^T$	$\mathbf{t}^T$	$\mathbf{b}^T$	$\mathbf{n}^T$	$\frac{d\mathbf{g}^T}{dL_u}$	$\frac{d^2\mathbf{g}^T}{dL_u^2}$	$\frac{d\mathbf{t}^T}{dL_u}$	$\frac{d\mathbf{b}^T}{dL_u}$	$\frac{d\mathbf{n}^T}{dL_u}$	$\frac{d^2\mathbf{t}^T}{dL_u^2}$	$\frac{d^2\mathbf{b}^T}{dL_u^2}$	$\frac{d^2\mathbf{n}^T}{dL_u^2}$
0	...	...	...	...	...	...	...	...	...	...	...	...
ΔL_u	...	...	...	...	...	...	...	...	...	...	...	...
...	...	...	...	...	...	...	...	...	...	...	...	...
$n_u\Delta L_u$	...	...	...	...	...	...	...	...	...	...	...	...

Fig. 1.3 Structure of a look-up-table of a spatial curve geometry for application on the definition of path-following kinematic constraints or in contact

L_v	0				ΔL_u					$n_u\Delta L_u$			
L_u	$\mathbf{g}^T$	$\mathbf{t}^T$	$\mathbf{b}^T$	$\mathbf{n}^T$	$\mathbf{g}^T$	$\mathbf{t}^T$	$\mathbf{b}^T$	$\mathbf{n}^T$		$\mathbf{g}^T$	$\mathbf{t}^T$	$\mathbf{b}^T$	$\mathbf{n}^T$
0	...	...	...	...	...	...	...	...	...	...	...	...	...
ΔL_u	...	...	...	...	...	...	...	...		...	...	...	...
...	...	...	...	...	...	...	...	...		...	...	...	...
$n_u\Delta L_u$	...	...	...	...	...	...	...	...		...	...	...	...

Fig. 1.4 Structure of a look-up-table of a freeform surface geometry for application in contact

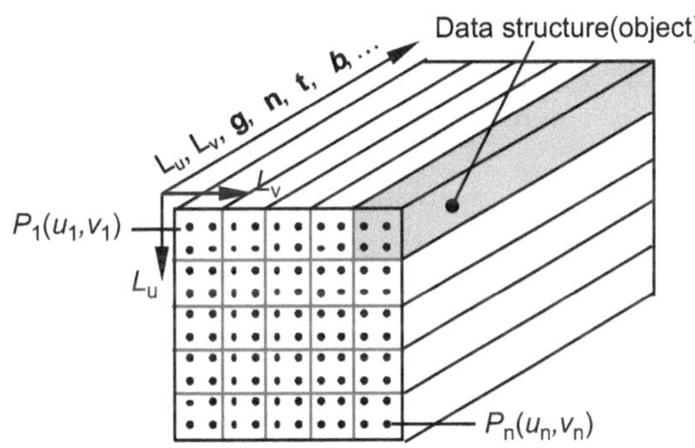

Fig. 1.5 Schematic representation of the lookup table of a surface

or large surfaces, with small arc length steps the size of the table becomes too large to keep in memory. The first problem is easily addressed when using objects, or structures, in the computational implementation of the LuT, as implied in Fig. 1.5. Each element (i,j) of the structure, with the dimension of $n_u \times n_v$, contains the arc-length parameters L_u and L_v and vectors $\mathbf{g}$, with the point position in the surface reference frame, $\mathbf{t}$, $\mathbf{b}$ and $\mathbf{n}$. If the LuT is to be used in the definition of kinematic constraints, the parametric derivatives of the point position and vectors may also be necessary [66].

A bilinear interpolation of the twelve point coordinates and vector components is performed in each instant of simulation in order to evaluate all the relevant geometric information for the contact detection process [43, 66]. Note that the accuracy of the interpolation is solely dependent on the arc-length step of the table,

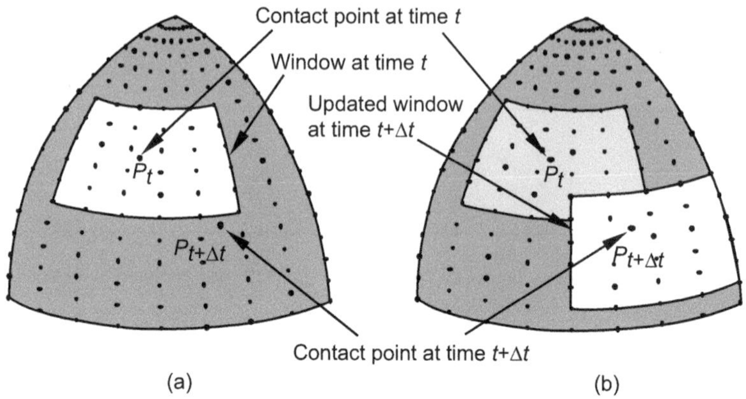

Fig. 1.6 Representation of an update of a storage window for: (**a**) Time step t; Time step $t+\Delta t$

being this chosen such a way that a pre-defined maximum interpolation error is never exceeded.

The partial reading of a surface file is suitable for contact point searching, in particular when the surface file is very large or when different rigid bodies share common geometric features. In these cases, the storage window must contain the contact zone, i.e., it should contain the records to which the candidate contact points belong. Figure 1.6a shows one eighth of a spherical surface, previously illustrated in Fig. 1.2b, with a storage window that includes the contact point at the instant of time t. If at the next instant of time, $t+\Delta t$, the contact point remains in a region close to the previous contact point, the storage window does not need to be updated. In contrast, when the next contact point approaches the border of the current window, the storage window has to be updated such that the contact point is centered in the new window, as in Fig. 1.6b.

The need for updating the storage window is checked in each contact calculation. This process is summarized by the following steps:

1. Locate contact record, i.e., the record that contains the initial guess (u_0,v_0);
2. Check if contact record belongs to storage window. If not, go to step (4);
3. Check if contact record is a border record. If not, go to step (7);
4. Locate storage window considering the contact record the center of the new storage window;
5. Store the index of the records of the new storage window, namely the first record and the border records;
6. Read and store the data of the records that belong to the new window;
7. Proceed with the contact computation.

The possibility to read partially the surface is useful in the cases where the contact point moves slightly and smoothly and, therefore, it remains in the vicinity of the previous contact point. Thus, the amount of memory used is significantly reduced and the contact detection process is more efficient.

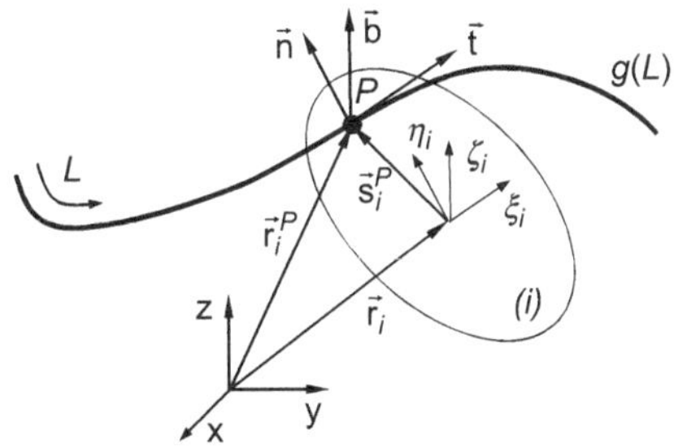

Fig. 1.7 Prescribed motion constraint, enforcing a body to follow a curve and its orientation to follow that of the curve moving frame

1.2.3 *Kinematic Constraints with Spatial Curves*

The definition of motion prescribed constraints along a spatial curve are used in cams, path-following joints or in roller coasters, among others. To define a kinematic joint in which the motion of a body is constrained to follow a spatial curve, consider a point P, located on a rigid body i constrained to follow a path, as in Fig. 1.7. The path is defined by a parametric curve $\mathbf{g}(L_u)$, in which is parameter L_u represents the length travelled by the point along the curve from the origin to its current location. The constraint equations that enforce point P to follow the reference path $\mathbf{g}(L_u)$ are written as [43]:

$$\boldsymbol{\Phi}^{(pmc,3)} = \mathbf{0} \quad \equiv \quad \mathbf{r}_i^P - \mathbf{g}\left(L_u\right) = \mathbf{0} \tag{1.16}$$

where $\mathbf{r}_i^P$ is the position vector associated to point P, depicted in Fig. 1.7.

The prescribed motion constraint also ensures that the spatial orientation of body i remains unchanged with respect to the moving Frenet frame (**t**,**n**,**b**) associated to the curve. Consider a rigid body i where $(\mathbf{u}_\xi,\ \mathbf{u}_\eta,\ \mathbf{u}_\zeta)_i$ represent the unit vectors associated to the axes of $(\xi,\eta,\zeta)_i$ defined in the body frame. Consider also that the Frenet frame of the general parametric curve $\mathbf{g}(L)$ is defined by the principal unit vectors $(\mathbf{t},\ \mathbf{n},\ \mathbf{b})_L$, as depicted in Fig. 1.7. The relative orientation between the body vectors $(\mathbf{u}_\xi,\mathbf{u}_\eta,\mathbf{u}_\zeta)_i$ and the curve local frame $(\mathbf{t},\ \mathbf{n},\ \mathbf{b})_L$ must be such that [43]:

$$\boldsymbol{\Phi}^{(lfac,3)} = \begin{Bmatrix} \mathbf{n}^T\mathbf{A}_i\mathbf{u}_\xi \\ \mathbf{b}^T\mathbf{A}_i\mathbf{u}_\xi \\ \mathbf{n}^T\mathbf{A}_i\mathbf{u}_\eta \end{Bmatrix} - \begin{Bmatrix} a \\ b \\ c \end{Bmatrix} = 0 \tag{1.17}$$

where $\{a \quad b \quad c\}^T = \text{diag}\left[\left(\mathbf{A}_L^0\right)^T \mathbf{A}_i^0\right]$ are constants calculated at the initial time of the analysis by using Eq. (1.17) with the initial conditions. The contribution of the prescribed motion constraint to the constraint acceleration is written as

$$\begin{bmatrix} \boldsymbol{\Phi}_\mathbf{q}^{(pmc,3)} \\ \boldsymbol{\Phi}_\mathbf{q}^{(lfac,3)} \end{bmatrix} \ddot{\mathbf{q}} = \begin{bmatrix} \boldsymbol{\gamma}^{(pmc,3)\#} \\ \boldsymbol{\gamma}^{(lfac,3)\#} \end{bmatrix} \tag{1.18}$$

where the Jacobian matrix associated to each part of the kinematic constraint is

$$\mathbf{\Phi}_{\mathbf{q}}^{(pmc,3)} = \begin{bmatrix} \mathbf{I} & -\tilde{\mathbf{s}}_i^R \mathbf{A}_i & -d\mathbf{g}/dL \end{bmatrix} \tag{1.19}$$

$$\mathbf{\Phi}_{\mathbf{q}}^{(lfac,3)} = \begin{bmatrix} \mathbf{0}^T & - & \mathbf{n}^T \mathbf{A}_i \tilde{\mathbf{u}}_\xi & (d\mathbf{n}/dL)^T \mathbf{A}_i \mathbf{u}_\xi \\ \mathbf{0}^T & - & \mathbf{b}^T \mathbf{A}_i \tilde{\mathbf{u}}_\xi & (d\mathbf{b}/dL)^T \mathbf{A}_i \mathbf{u}_\xi \\ \mathbf{0}^T & - & \mathbf{n}^T \mathbf{A}_i \tilde{\mathbf{u}}_\zeta & (d\mathbf{n}/dL)^T \mathbf{A}_i \mathbf{u}_\zeta \end{bmatrix} \tag{1.20}$$

and the contribution of each part of the kinematic constraint to the right hand side of the acceleration equations is

$$\gamma^{(pmc,3)\#} = -\tilde{\boldsymbol{\omega}}_i \, \dot{\mathbf{s}}_i^R + \frac{d^2\mathbf{g}}{dL^2} \dot{L}^2 \tag{1.21}$$

$$\gamma^{(lfac,3)\#} = -\left\{ \begin{matrix} \left[2\,\dot{L}(d\mathbf{n}/d\mathrm{L})^T \mathbf{A}_i \tilde{\boldsymbol{\omega}}'_i + \mathbf{n}^T \mathbf{A}_i \tilde{\boldsymbol{\omega}}'_i \tilde{\boldsymbol{\omega}}'_i + \dot{L}^2 \left(d^2\mathbf{n}/dL^2\right)^T \mathbf{A}_i \right] \mathbf{u}_\xi \\ \left[2\,\dot{L}(d\mathbf{b}/dL)^T \mathbf{A}_i \tilde{\boldsymbol{\omega}}'_i + \mathbf{b}^T \mathbf{A}_i \tilde{\boldsymbol{\omega}}'_i \tilde{\boldsymbol{\omega}}'_i + \dot{L}^2 \left(d^2\mathbf{b}/dL^2\right)^T \mathbf{A}_i \right] \mathbf{u}_\xi \\ \left[2\,\dot{L}(d\mathbf{n}/dL)^T \mathbf{A}_i \tilde{\boldsymbol{\omega}}'_i + \mathbf{n}^T \mathbf{A}_i \tilde{\boldsymbol{\omega}}'_i \tilde{\boldsymbol{\omega}}'_i + \dot{L}^2 \left(d^2\mathbf{n}/dL^2\right)^T \mathbf{A}_i \right] \mathbf{u}_\zeta \end{matrix} \right\} \tag{1.22}$$

The minimum requirements for the degree of the interpolating polynomials that can be used in the formulation of the prescribed motion constraint is associated to the order of the derivatives used in Eqs. (1.16, 1.17, 1.18, 1.19, 1.20, 1.21 and 1.22). The right hand side vector in Eq. (1.21) involves $d^2\mathbf{n}/dL^2$ being $\mathbf{n} = \mathbf{k}/\|\mathbf{k}\|$ given by Eq. (1.9) and $\mathbf{k} = \mathbf{g}^{uu} - (\mathbf{g}^{uuT}\mathbf{g}^u)\mathbf{g}^u/\|\,\mathbf{g}^u\|^2$ by Eq. (1.10). Therefore, it is required that the fourth derivative of the interpolating polynomial is used, being a quintic polynomial the lowest odd degree polynomials to be used to formulate accurately the prescribed motion constraint [44].

1.2.4 Application in a Roller Coaster Dynamics Study

Roller coasters are, apparently, relatively simple mechanical systems when compared to modern railways or cars, but are a perfect examples for using spatial curve geometries. Since most roller coasters represent unique designs, extensive testing of real world prototypes is not possible from an economical point of view, being computational tools of important for virtual testing before the construction of the roller coaster track. The geometric description of the curve for a roller coaster track starts with the definition of a moving referential, being the track used in this application depicted by the sweep of the normal and binormal vectors shown in Fig. 1.8. The individual geometry of each rail of the roller coaster track is obtained from the definition of the centerline and track torsion using the moving frames

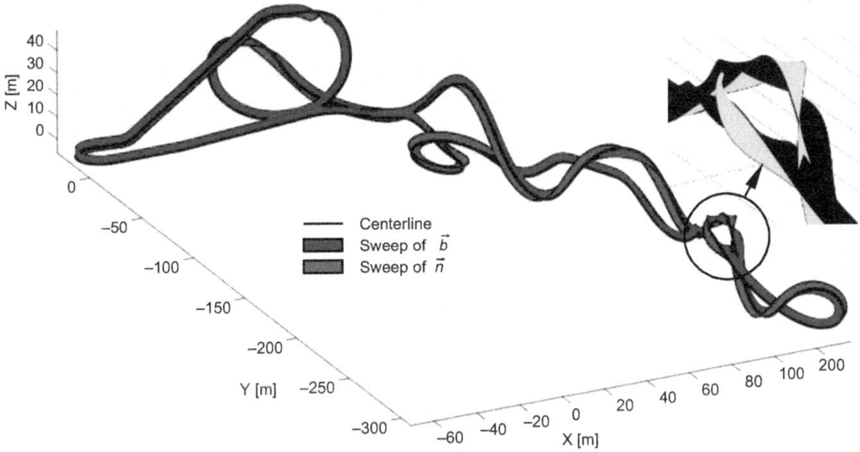

Fig. 1.8 Roller coaster track with the identification of the sweeps of the normal and binormal vectors of the moving frame, highlighting a corkscrew segment of the track

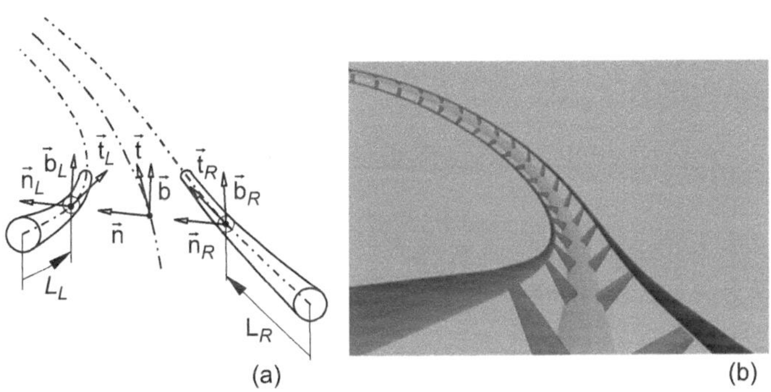

Fig. 1.9 Spatial description of the roller coaster rails: (**a**) Frenet and rails moving frames: (**b**) Geometry of the rails for a roller coaster

shown in Fig. 1.9. Note that, in the definition of the moving referential there are singularities in the straight track segments, as discussed by Tandl and Kecskemethy [45, 46], that are handled by enforcing the normal vector to be parallel to the XY plane. Furthermore, due to the long track length, LuT are used in the implementation of the path following kinematic constraints

A model of a roller coaster vehicle and biomechanical multibody model are developed, and described in the work by Viegas et al. [67], being its dynamics analyzed with the objective of studying the occupant exposure to injurious conditions. The motion resulting from the simulation of the roller coaster is depicted in Fig. 1.10 with selected frames of the vehicle and passenger motion.

The track torsion is the rotation of the track tangent plane, about the tangent vector, with respect to the plane defined by the tangent and normal vectors of the

Fig. 1.10 Selected frames from the motion of a roller coaster vehicle with an occupant

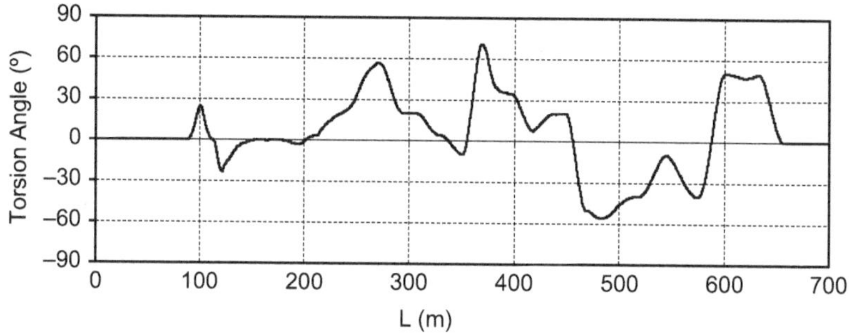

Fig. 1.11 Track torsion along its complete length

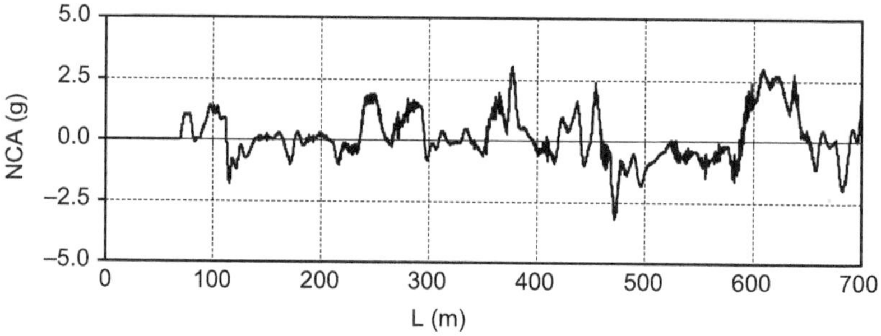

Fig. 1.12 Non-compensated acceleration of a passenger seating in the front left seat

Frenet frame associated to the track centerline, being its evolution for the roller coaster track considered here shown in Fig. 1.11. The design of the evolution of the torsion is used not only for some of the thrills of a roller coaster ride but also to allow controlling the lateral accelerations experienced by roller coaster passengers.

Among other results of interest from the dynamic analysis of the roller coaster, those associated to the rider safety are of importance, being the non-compensated acceleration, directly associated to the lateral accelerations experienced by the rider, shown in Fig. 1.12, one of the representative quantities.

The analysis of the non-compensated acceleration, and other biomechanical injury criteria, allows the designer to reshape the segments to avoid the passenger exposure to be close to any injurious condition. Another form of using the passenger data is in the redesign of the track to obtain the thrills and emotions desired by roller coaster users.

1.3 Contact Mechanics in Multibody Systems

Contact problems present some of the most important challenges of the correct and efficient definition of complex spatial geometries in multibody dynamics. Contact mechanics involves the solution of two basic sub-problems: the contact detection and the contact force modelling. In this work the contact identification problem is analyzed and continuous contact models, based on Hertzian and non-Hertzian elastic contact are considered. Nonsmooth mechanics approaches are not considered here, although they offer good alternatives to handle contact mechanics [51, 57–60, 68]. The numerical issues associated with the contact-impact mechanics are also overviewed in the process, receiving the implications of the contact force modelling on time integration attention [69, 70].

1.3.1 Search of Contact Between Two Generic Surfaces

Let it be assumed that a body in motion approaches another. These bodies can be described as generic surfaces, which can be represented as shown in Fig. 1.13. At any given instant, there are always two points which are closest to each other and where contact is more likely to occur. The search for contact between the two surfaces requires the identification of two parameters of each surface, associated to the location of the points that are either in contact or in closer proximity. This is illustrated in Fig. 1.13a, in which the points in closer proximity are depicted as P and Q, each belonging to a generic surface on body i and j, respectively. The surfaces are defined in the referentials $(\xi\eta\zeta)_i$ and $(\xi\eta\zeta)_j$ fixed to the centers of mass of bodies i and j, respectively. Vector $\mathbf{d}$ represents the distance between them, given by $\mathbf{d} = \mathbf{r}_P - \mathbf{r}_Q$.

Still with reference to Fig. 1.13, on point P in body i the vector normal to the surface is $\mathbf{n}_P$ while $\mathbf{t}_P$ and $\mathbf{b}_P$ are the tangent and binormal vectors to the surface, respectively, forming an orthogonal basis. The same applies to $\mathbf{n}_Q$, $\mathbf{t}_Q$ and $\mathbf{b}_Q$, which are the vectors on point Q. The relation between the components of the different vectors defined in each body and the parameters used to define the surface depends on the specific geometry of the contacting surface. The conditions for minimal distance between the two surfaces are that points P and Q are generically described by [71],

$$\begin{cases} \mathbf{d}^T\ \mathbf{t}_Q = 0 \\ \mathbf{d}^T\ \mathbf{b}_Q = 0 \end{cases} \quad ; \quad \begin{cases} \mathbf{n}_Q^T\ \mathbf{t}_P = 0 \\ \mathbf{d}^T\ \mathbf{b}_P = 0 \end{cases} \tag{1.23}$$

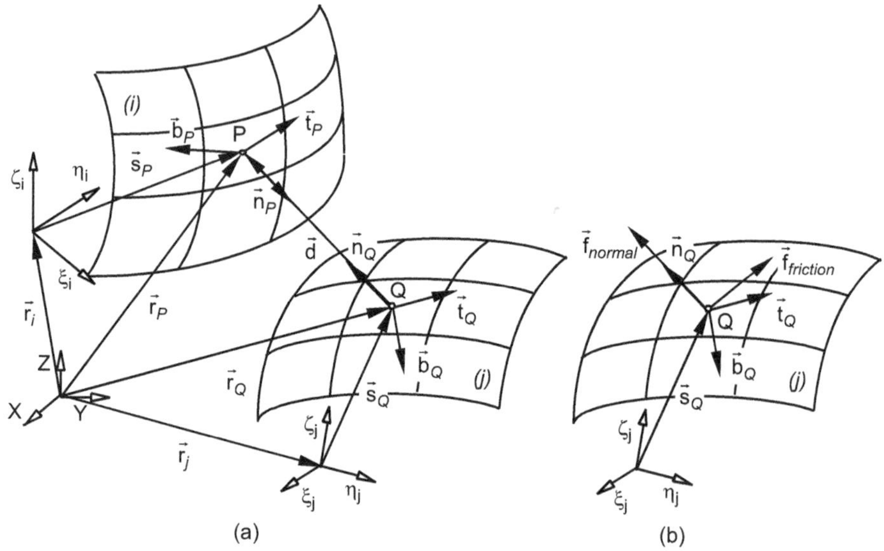

Fig. 1.13 Contact points candidates in two parametric surfaces. Point P belongs to a surface on body i and point Q to a surface on body j: (**a**) Contact detection; (**b**) Contact forces in body i

which means that not only the normal to each surface must be collinear with the vector that connects the two points in closer proximity but also perpendicular to the tangent and binormal vectors on each point. Effective contact occurs if, besides the fulfilment of Eq. (1.23), penetration also exists, which is expressed by,

$$\delta = \mathbf{d}^T \, \mathbf{n}_Q \leq 0 \tag{1.24}$$

otherwise, the points are in close proximity, but not in contact.

1.3.2 Simple Continuous Contact Force Models

When contact between two surfaces exists both normal and friction forces develop, as highlighted in Fig. 1.13b. Depending on the application, the constitutive equations for the frictions forces may be simple or very complex. Road vehicle dynamics, in which the tire to road contact forces are described by elaborate tire models [72, 73], railway dynamics, in which the wheel to rail contact lead to friction forces requiring a rather complex interaction [71, 74], or even in lubricated rotating machinery, in which the tribological tangential forces are associated to lubrication modes and specific lubrication force models [75], present examples of systems for which the tangential contact forces calculation require complex constitutive laws.

However, for many applications of multibody systems simpler interaction force models are devised, mostly based on Hertzian contact conditions.

The modified Kelvin-Voigt's visco-elastic contact model [47, 50] is a simple continuous normal contact force model that relates the interference between two surfaces with a normal reaction force, formulated as,

$$f_{normal} = K\,\delta \begin{cases} c_e & \dot{\delta} \leq 0 \\ \left[c_e + (1 - c_e)\left(3r^2 - 2r^3\right)\right] & 0 < \dot{\delta} < v_0 \\ 1 & \dot{\delta} \geq v_0 \end{cases} \tag{1.25}$$

with the ratio between the penetration velocity and the penetration velocity tolerance being $r = \dot{\delta}/v_0$, a penalty stiffness K and a coefficient of restitution c_e. The penetration velocity tolerance, pre-defined by the user as being $v_0 = 0.1$ m/s is often used. Note that the modified Kelvin-Voigt normal contact force model includes the penetration velocity tolerance to smooth the discontinuity in the contact force in the transition between compression and restitution but also includes an energy dissipation term during the restitution contact phase.

The normal contact force model proposed by Lankarani and Nikravesh [76] is based on the assumption for Hertzian contact and also includes a dissipative term based on the Hunt-Crossley model [77]. This normal force contact model is suitable for low impact velocities in which local plasticity effects do not develop or are negligible,

$$f_{normal} = K\,\delta^n \left[1 + \frac{3\left(1 - c_e^2\right)}{4}\frac{\dot{\delta}}{\dot{\delta}^{(-)}}\right] \tag{1.26}$$

where δ is evaluated using Eq. (1.24), and includes the relative stiffness of the contacting surfaces K, the pseudo-penetration exponent n and the restitution coefficient c_e dependent on the geometry and material of the contacting surfaces. $\dot{\delta}$ is the velocity of indentation and $\dot{\delta}^{(-)}$ is the velocity of indentation at the initial instant of contact. Note that ratio $\delta/\dot{\delta}^{(-)}$ leads to numerical problems for very small velocities of indentation at the start of contact, i.e., when $\dot{\delta}^{(-)} \approx 0$. In the computational implementation of the Lankarani and Nikravesh model, or any other that has the dissipative part written in the same form, the ratio $\delta/\dot{\delta}^{(-)} = 1$ when the penetration velocity $\dot{\delta}$ exceeds $\dot{\delta}^{(-)}$ or when $\dot{\delta}^{(-)} \approx 0$.

The contact between two surfaces not only generates normal forces but also friction forces, if the relative tangential velocity between the surfaces is not null. The Amontons-Coulomb friction model is not only the most common form of describing the friction forces but also the basis for many other models. Here, this friction force model is modified to include a regularization term to prevent numerical problems during integration of the system equations of motion, being [47].

$$f_{friction} = -c_f c_d f_{normal} \dot{\delta}/\left|\dot{\delta}\right| \tag{1.27}$$

where c_f is the dynamic friction coefficient, c_d a numerical regularization factor and $\dot{\delta}$ is the relative angular velocity of the bodies connected by the joint. The regularization factor proposed by Threfal is the exponential function $c_d = 1 - \exp\left(-3\dot{\delta}/v_1\right)$ while Ambrósio [78] proposes a ramp function defined as $c_d = 0$ for $\dot{\delta} \leq v_0$, $c_d = \left(\dot{\delta} - v_0\right) / \left(v_1 - v_0\right)$ for $v_1 \leq \dot{\delta} \leq v_0$ and $c_d = 1$ otherwise. Therefore, the input parameters for these models are the dynamic friction coefficients c_f, the threshold velocities v_0 and v_1 and the joint friction tuning normal force f_n.

In a wide number of contact problems, the transition between static and dynamic friction plays an important role. The Stribeck effect aims at representing numerically such transition with continuous and smooth numerical description. There are several friction models, which include the Stribeck effect to represent static friction, proposed [79]. Among these, the Bengisu and Akay model, with good numerical characteristics to be used in contact problems, is defined by [80]

$$n_{friction} = -f_{normal}\dot{\delta}/\left|\dot{\delta}\right| \begin{cases} \frac{c_s}{v_0^2}\left(\left|\dot{\delta}\right| - v_0\right)^2 + c_s & \left|\dot{\delta}\right| \leq v_0 \\ c_f + \left(c_f - c_s\right) e^{-\xi\left(\left|\dot{\delta}\right| - v_0\right)} & \left|\dot{\delta}\right| > v_0 \end{cases} \tag{1.28}$$

where v_0 is the Stribeck velocity, c_s is the static friction coefficient and ξ is the curve shape factor, typically in the order of to 50 s/rad.

The use of friction models that include the Stribeck effect, such as the Bengisu and Akay model, have the advantage of representing the transition from static to dynamic friction better than those described by Eq. (1.27). Other alternative friction models, are designed to better capture the pre-sliding displacement and the frictional lag, being designated as dynamic friction models [79]. Among these the Gonthier friction model, which is an evolution of the LuGre friction model, shows good numerical properties being suitable for applications in which the stick-slip transitions play important roles in the system dynamics. Due to their complexity they are not overseen in this work being the interested reader referred to the work by Marques et al. [79].

1.3.3 *Numerical Aspects of the Contact Analysis*

One of the critical aspects in the dynamic simulation of the multibody systems with collisions is the detection of the precise instant of contact. The contact duration and the penetration cannot be predicted from the pre-impact conditions due to the influence of the kinematic constraints imposed in the bodies on the overall system motion. Thus, before the first impact, the bodies can freely move relative to each other and, in this phase, the step size of the integration algorithm may become relatively large. Therefore, if the numerical integration is not handled properly, the first impact between the colliding bodies is often made with a high penetration depth, and, hence, the calculated contact forces becomes artificially large.

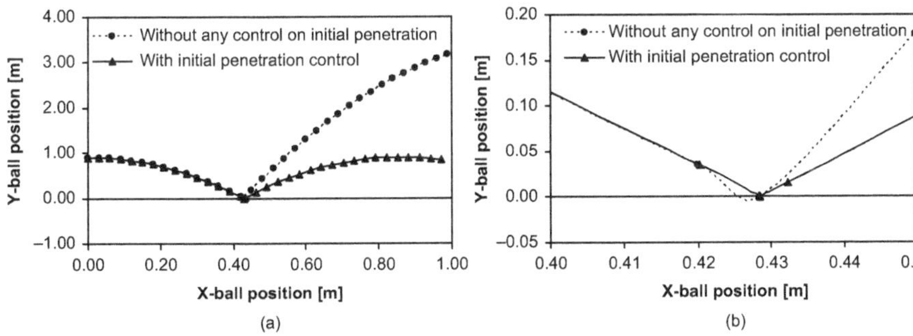

Fig. 1.14 (**a**) Trajectory of a falling ball obtained with integration algorithms with and without initial penetration control; (**b**) Detailed view in the vicinity of contact

The importance of the initial penetration control, in the framework of the integration of the equations of motion, is better discussed using a simple example [69]. Take the case of the falling ball illustrated in Fig. 1.14, animated by an initial horizontal velocity and acted upon by gravity forces only. The motion of the ball is such that during its falling trajectory it strikes the ground. The penetration of the ball in the ground, in the integration time step, for which contact is first detected, is,

$$\delta^{(-)} = (h - R) - y_b \tag{1.29}$$

where y_b is the y coordinate of the ball center of mass. The superscript $(-)$ on δ means that it is the penetration when contact is first detected. Note that $\delta^{(-)}$ must be a positive value for contact. Therefore, by monitoring the sign of the penetration at every time step $t+\Delta t$ the start can be identified from,

$$\delta^{(-)}(\mathbf{q}, t)\,\delta^{(-)}(\mathbf{q}, t + \Delta t) \leq 0 \tag{1.30}$$

When Eq. (1.30) is fulfilled the start of contact occurs at $t+\Delta t$. The integration of the equations of motion of the system can proceed with no numerical problem if the penetration first detected is close to zero, or at least below a pre-defined threshold, i.e., if $\delta^{(-)}(\mathbf{q}, t + \Delta t) \leq \delta_{max}$. Because this is not always the case, strategies to limit the time step in the vicinity of contact must be implemented when solving contact problems [69].

Define as δ^{-} the distance between the two surfaces in the time step t^{-} that precedes the time step t^{+}, at which penetration δ^{+} is first detected. In between these time steps, say at t^{c}, the penetration $\delta^{c} = 0$ exists. Assuming constant velocity for the multibody system in the vicinity of contact, the time at which contact starts can be calculated by,

$$t^{c} = t^{-} + \frac{\delta^{-}}{\delta^{+} - \delta^{-}}\Delta t \tag{1.31}$$

Consequently, the ideal situation, during the integration of the multibody equations of motion, would be that a time step in the vicinity of contact to be

$$\Delta t^{ideal} = t^c - t^- + \varepsilon \tag{1.32}$$

where ε is a very small number to effectively ensure that $\delta_{max} > \delta^c > 0$. Several procedures are suggested to ensure that $\delta^+ < \delta_{max}$, can be implemented, depending on the access that exists to the numerical integrator. Several strategies are proposed by Flores et al. [69] to handle the numerics associated to the regularization of the time integration of the multibody equations of motion, being the interested reader referred to their work.

1.3.4 *Selected Applications*

The applications explored here for contact between bodies with a prescribed geometry are intended to identify and propose solutions for pitfalls in the contact detection and to present realistic application cases with rather complex contact mechanics. The solution of the contact between simple surfaces, circles in a planar contact case and cylinders in a spatial case, explore the pitfalls on the contact detection and propose measures to avoid them. The wheel-rail contact problem, associated to railway dynamics, presents a more complex, and realistic, contact problem for which the procedures outlined in this work are of fundamental importance.

1.3.4.1 Circle-to-Circle and Cylinder-to-Cylinder Contact Detection

The contact identification problem of two circles in a two-dimensional space is used here not only to test the contact detection procedures presented but also to identify pitfalls of the numerical methods involved. According to Fig. 1.15a, the local position of the potential contact points in the surfaces are $\mathbf{s}'^P_i = R_j\{\cos\varphi_j \quad \sin\varphi_j\}^T$ and $\mathbf{s}'^Q_i = R_i\{\cos\varphi_i \quad \sin\varphi_i\}^T$, while the normal and tangential vectors associated to the surfaces on those points are $\mathbf{n}_P = \mathbf{A}_j\{\cos\varphi_j \ \sin\varphi_j\}^T$, $\mathbf{t}_P = \mathbf{A}_j\{\sin\varphi_j \ -\cos\varphi_j\}^T$, $\mathbf{n}_Q = \mathbf{A}_i\{\cos\varphi_i \ \sin\varphi_i\}^T$ and $\mathbf{t}_Q = \mathbf{A}_i\{\sin\varphi_i \ -\cos\varphi_i\}^T$, being R_i and R_j the circle radios. The conditions for minimal distance are defined, in this planar case, as:

$$\begin{cases} \mathbf{n}_j^T \mathbf{t}_i = 0 \\ \mathbf{d}^T \mathbf{t}_i = 0 \end{cases} \tag{1.33}$$

which is solved using any solver for systems of nonlinear equations, such as the Newton-Raphson method, optimization based procedures or other suitable approach [81].

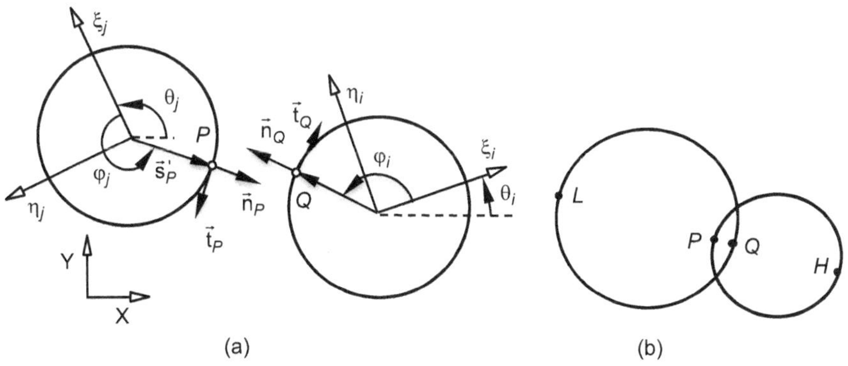

Fig. 1.15 Geometric relations describing the positions of points P and Q in two circles; (**a**) Geometric definitions for the contact problem; (**b**) Potential contact points to form the contact pairs

Table 1.1 Results from contact point detection for two random circles

Matlab method	Success rate (%)	Computer time (s)
`Newton-Raphson`	66.7	0.10
`fsolve`	100	0.15
`fmincon`	100	0.37

This system of nonlinear equations has four solutions for potential contact points, depicted as the pairs of points *PQ*, *HP*, *QL* and *LH*. Although, by visual inspection of Fig. 1.15b, it is clear that *PQ* represent the contact pair there is no reason for the solution of the system of nonlinear equations converge for this pair and not to any of the others. To minimize the problem, a discrete method for finding a close guess to the solution is necessary so that the solver is guided to the solution closer to the guess. However, there is no assurance that another solution is not chosen instead.

To test the different methods to solve the nonlinear system of equations 100 pairs of circles are generated in random positions and orientations. With a Matlab® code in which the functions `NewtonRaphson`, `fsolve` and `fmincon` are used and the initial guess for the location of each pair of points in closest proximity are offered as the first guess for the solution to obtain a solution. By visual inspection of each of the solution found, the result is deemed successful or not. Table 1.1 presents the success rate of each of the methods tested as well as the computer time required to solve the 100 problems with each method tested.

This analysis clearly shows that Newton-Raphson is not always capable of finding the correct result in a reliable way, being the wrong solution always associated with contact pairs that exhibit a very large interference. Since in most, if not all, contact problems using rigid bodies the indentations are very small a limitation on the maximum indentation of 1% of the largest circle radius is set in the random generation of the 100 circle pairs. It is observed that for this new population the Newton-Raphson method has a success rate of 100% in finding the correct contact pair.

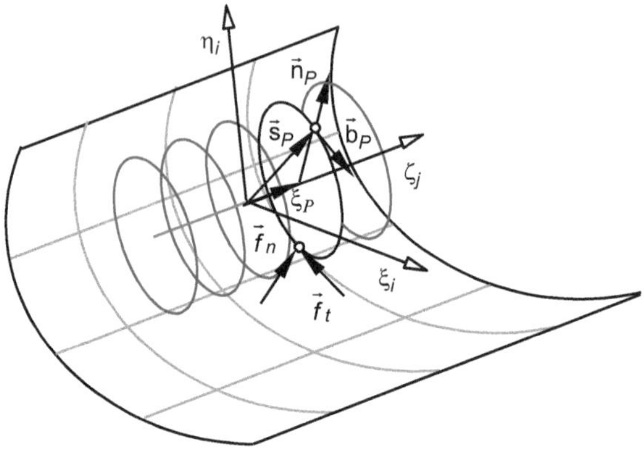

Fig. 1.16 Slice method applied to a cylindrical roller

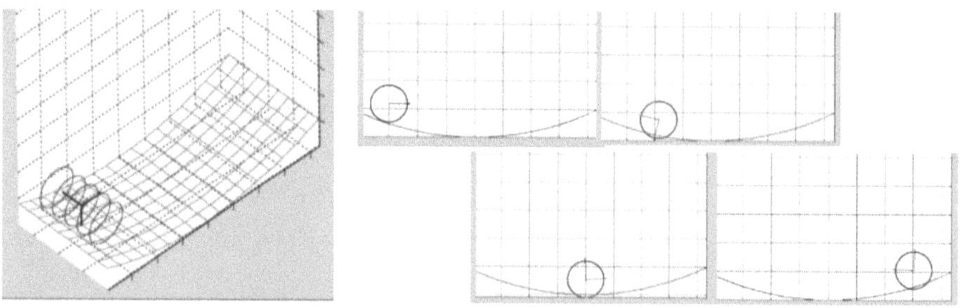

Fig. 1.17 3D view of the starting point for a roller on an internal cylindrical surface and its motion observed in a side view of the roller and cylindrical surfaces

An approach to model contact problems with spatial surfaces of revolution is the strip method, in which a revolution surface is discretized by a finite number of circles. Now, the contact of each individual circle with a surface, as shown in Fig. 1.16, is solved independently. In this form, the line contact associated to the contact pairs is discretized in a series of point contacts. This approach is used in roller bearing dynamics for the contact between the rollers and the raceways and cage [82] and in railway dynamics for cases in which the wheel profile leads to conformal wheel-rail contact [83].

The dynamics of the contact of the roller, using the slice discretization, on a cylindrical surface is computed for a period of time that allows the roller to travel for the complete arc-length of the cylinder, as shown in Fig. 1.17. The Kelvin-Voigt normal contact force model, described by Eq. 1.25, and the Amontons-Coulomb friction force model with Threfal smoothing, described by Eq. (1.27), are used in the contact.

A large number of initial conditions and material properties are simulated, each one of them using `Newton-Raphson`, `fsolve` and `fmincon` methods. It is observed that all methods used are successful in finding the correct solution of all contacts involved in the dynamics of the test case. However the computational time required by each method is quite different, being `Newton-Raphson` always faster, `fsolve` about 1.5 times slower and `fmincon` about 4 times slower. The implementation of the Newton-Raphson method is more demanding than the other methods, as it requires that the Jacobian matrices for each contact pair are available. When dealing with several thousand of potential contact pairs in each time the equations of motion of the multibody system are solved, which may take place more than once per time step, as in the case of the dynamic analysis of roller bearings, computational efficiency becomes of major importance.

1.3.4.2 Wheel-Rail Interaction in Railway Dynamics

Railway dynamics involves, in general, the solution of the wheel-rail contact problem, which poses interesting challenges in terms of the contact detection but also on the modelling of the contact forces. From the point of view of the contact detection, the surface of revolution of the wheel may be convex or non-convex depending on its wear. The case presented here to illustrate the close relation between the contact problem and the geometrical description of the contact surfaces involves the operation of a light rail vehicle in an existing railway track.

The light rail vehicle, shown in Fig. 1.18, is composed by a carbody, for passengers, supported by two bogies. One bogie is powered by a diesel engine, the motor bogie, while the other one, the trailer bogie, is free. Each bogie is composed by a bogie frame, two wheelsets and four axleboxes, which include a primary suspensions. The carbody and bogie frames are connected by a secondary suspensions. Details of the railway vehicle multibody model can be found in the work by Magalhães et al. [84].

The vehicle operation is studied for a real railway track for which both for the horizontal nominal geometry, defined during the project, and for the current geometry, measured experimentally, are described in Fig. 1.19. Besides the curvature of track, as function of the track length, also the cant, the vertical profile and the track irregularities, all function of the track arc-length, are available and used for the study.

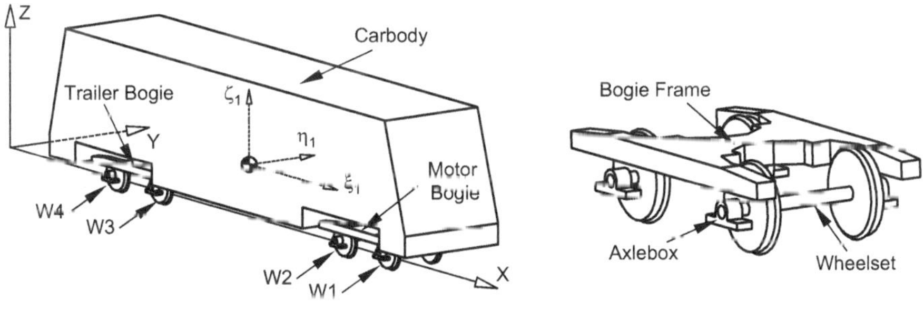

Fig. 1.18 Light Rail Vehicle multibody model with a more detailed bogie and wheelsets assembly

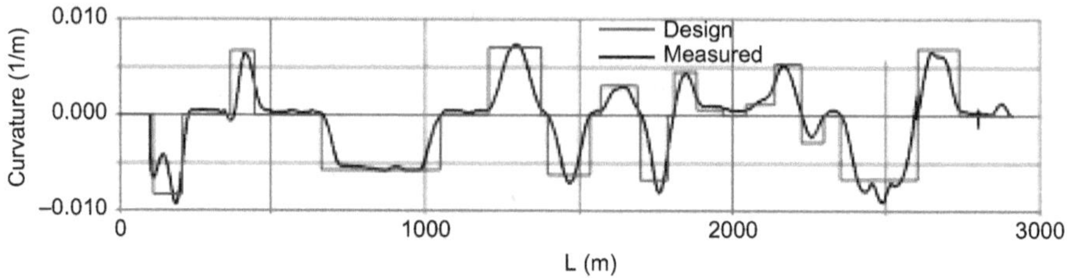

Fig. 1.19 Track curvature as function of the track arc-length for the measured and nominal data

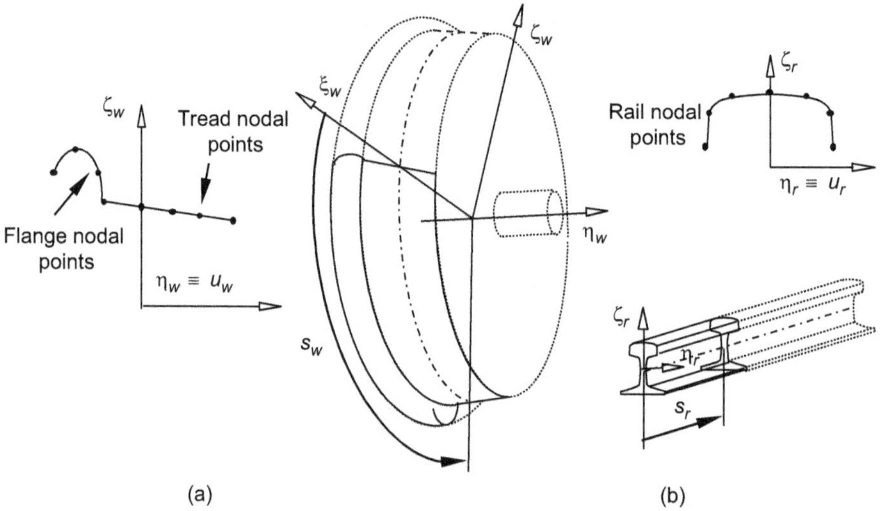

Fig. 1.20 Geometry of the bodies; (**a**) Wheel obtained as a surface of revolution of the wheel profile; (**b**) Rail surface obtained by sweeping the rail profile along the rail centerline

The surfaces of the rails, deemed as left and right rails, and the wheels, eight in each vehicle, are shown in Fig. 1.20 as sweep surfaces and surfaces of revolution, respectively. In the case of the wheel, there are, in fact, two surfaces of revolution, one for the tread and another for the flange. Therefore, the wheel rail contact between geometries with nominal configurations, i.e., geometries that are not worn, involves the solution of two independent problems per wheel-rail pairs: flange-to rail- and tread-to-rail contact [71].

Each one of the contact problems for each one of the railway vehicle wheels is solved using Eq. (1.23), leading to the identification of the contact points or those of closer proximity, as pictured in Fig. 1.21. This means that for each vehicle, equipped with eight wheels, sixteen contact problems need to be solved each time the equations of motion of the multibody system are solved. When contact is detected in each one of the contact pairs a normal contact force is evaluated using a contact model such as those described by Eqs. (1.25) or (1.26) and the creep, or tangential, forces are computed with an appropriate model, such as those proposed by Kalker [74], Polach [85], Chollet [86], Berg [87], Bruni [88], or similar.

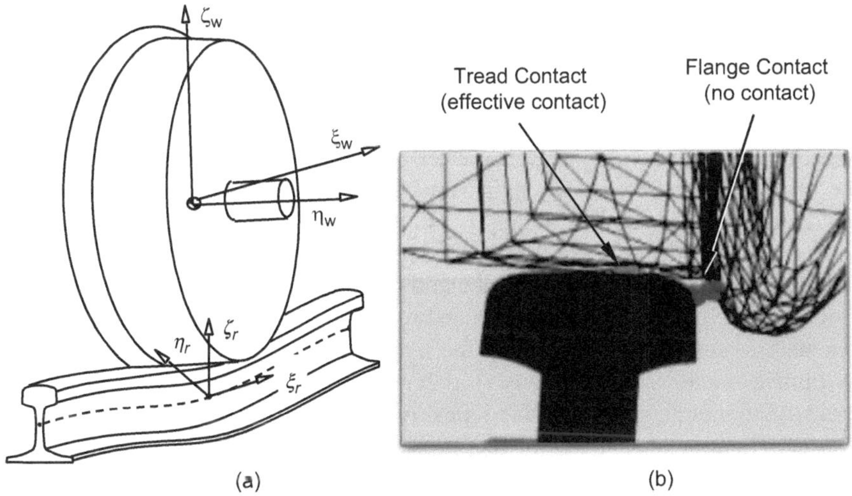

Fig. 1.21 Contact between wheel and rail; (**a**) Perspective view of the wheel-rail; (**b**) Identification of the independent treat-rail and flange rail contacts, where only the first is in effective contact

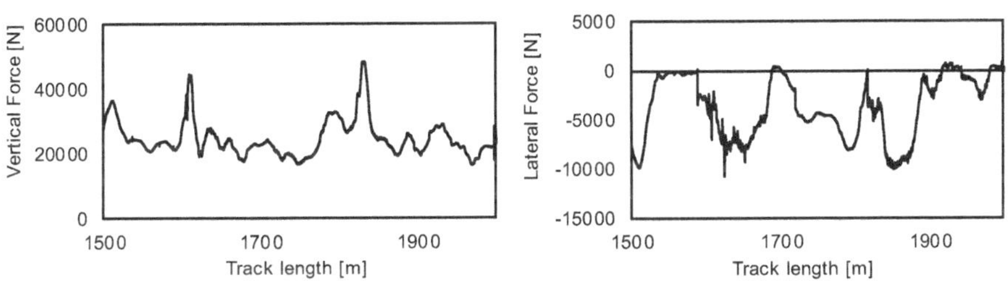

Fig. 1.22 Evolution of the wheel-rail contact forces along a section of the railway track

The rail-wheel contact force experienced by the vehicle wheels, for the scenario envisaged in this work, in which the vehicle is operating at a speed of 50 km/h, are illustrated by the normal and lateral, creep, forces shown in Fig. 1.22. It is noticeable the relation between the evolution of the lateral forces and the curves of the track.

The solution of the contact problem with non-worn wheels can be done online, as the work by Pombo et al. demonstrates [71]. However, it is also been demonstrated by some authors that computational efficiency is obtained from using LuT to identify the contact forces that develop in each contact pair [89]. In the case of worn, but still functional, wheel profile a conformal contact with the rail, for which the Hertzian contact force models do not apply, needs to be formulated. In this case, the wheel-rail contact is not only in the vicinity of a point but, eventually, in a larger region, being modelled as a multiple contact point problem. In these contact conditions the use of LuT to obtain the contact forces and the location of the equivalent contact points is unreplaceable, at least currently.

1.4 Perfect Kinematic Joints Versus Clearance Contact Joints

In a wide number of practical applications in multibody dynamics the kinematic joints are not perfect, exhibiting clearances and including elastomers to accommodate for misalignments and to damp vibrations. The inexistence of such features would not only impair the durability of the system but also prevent it from functioning properly [90–92]. The way of handling imperfect joints in multibody models is, instead of enforcing kinematic constraints between rigid bodies, by associating the relative displacements and rotations between the connected bodies to contact forces, i.e., by penalizing the relative displacements using appropriate constitutive relations due to contact, as for the clearance joints, or due to local elastic deformations, as for bushing joints. An overview of the general formulation proposed by Ambrosio and Pombo [50] is presented here, being the interested reader referred to the original reference for a detailed description of the methodology.

1.4.1 A General Formulation of a Clearance/Bushing Joint

A clearance/bushing joint constraining the relative motion between connected bodies requires the force penalization of the axial and radial displacements, axial misalignment and axial rotation described in Fig. 1.23. The joint reaction forces, here called also penalization forces, are applied in the bodies connected by the joint in the attachment points. The joint reaction forces are represented here as contact

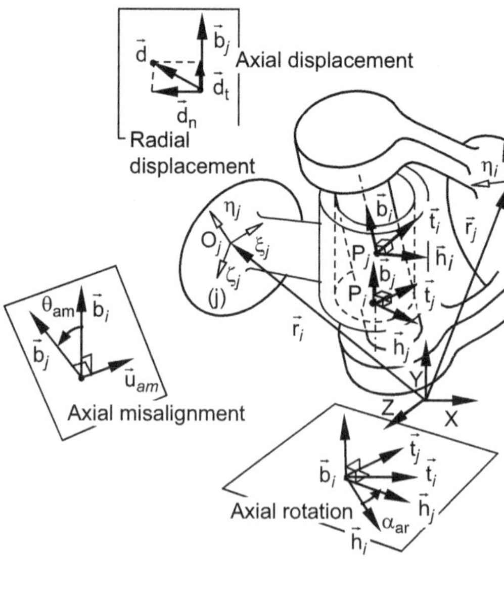

Fig. 1.23 Representation of a general clearance/bushing joint with the identification of specific relative motions required for its formulation

forces between the two bodies, using a methodology in line with those proposed by Ambrosio and Verissimo [48], for pure bushing joints, and by Flores et al. [47], for clearance joints.

To define the axial and radial displacements, axial misalignment and axial rotation, appearing in Fig. 1.23 let the distance vector **d** be decomposed into an axial displacement along vector $\mathbf{b}_i$, which specifies the joint axis in body j, defined as

$$\mathbf{d}_t = \left(\mathbf{d}^T\mathbf{b}_i\right)\mathbf{b}_i \tag{1.34}$$

and an orthogonal component, the radial displacement with respect to the joint axis, as

$$\mathbf{d}_n = \mathbf{d} - \mathbf{d}_t \tag{1.35}$$

The radial displacement, δ_n, and the radial direction, $\mathbf{u}_{rd}$, are

$$\delta_n = \sqrt{\mathbf{d}_n^T\mathbf{d}_n} \text{ and } \mathbf{u}_{rd} = \mathbf{d}_n/\delta_n \tag{1.36}$$

The axial displacement, δ_t, and the radial direction vector, $\mathbf{u}_{ad}$, are

$$\delta_t = \sqrt{\mathbf{d}_t^T\mathbf{d}_t} \text{ and } \mathbf{u}_{ad} = \mathbf{d}_t/\delta_t \tag{1.37}$$

The axial misalignment of the joint axis in bodies i and j is described by an angle θ_{am} measured about a vector $\mathbf{u}_{am}$ such that

$$\theta_{am} = \arcsin\left(\left|\tilde{\mathbf{b}}_i\,\mathbf{b}_j\right|\right) \text{ and } \mathbf{u}_{am} = \tilde{\mathbf{b}}_i\,\mathbf{b}_j/\left|\tilde{\mathbf{b}}_i\,\mathbf{b}_j\right| \tag{1.38}$$

The skew-symmetric matrix of a vector **b** is denoted by $\tilde{\mathbf{b}}$. If the axial misalignment is null, i.e., $\left|\tilde{\mathbf{b}}_i\,\mathbf{b}_j\right| \approx 0$, Eq. (1.38) is not required because the conditions of alignment of the axis of the bearing and journal are fulfilled. The axial rotation of the bearing, with respect to the journal, is defined by the angle

$$\alpha_{ar} = arc\ \sin\left(\mathbf{h}_i^T\mathbf{t}_j\right) \tag{1.39}$$

The evaluation of the misalignments and relative displacements between bearing and journal involve the same vectors used to setup the perfect kinematic joints. From the point of view of the model construction this means that the same is used to setup perfect kinematic and clearance/bushing joints. The cylindrical clearance/bushing joint is obtained by penalizing of the radial displacement and the axial misalignment between bearing and journal by a penalization forces written as

$$\begin{aligned}\mathbf{f}_i^{(cyl)} &= \mathbf{f}_{cyl}\left(\delta_n, \dot{\delta}_n, \delta_t, \dot{\delta}_t, \theta_{am}, \dot{\theta}_{am}\right)\\ \mathbf{f}_j^{(cyl)} &= -\mathbf{f}_i^{(cyl)}\end{aligned} \tag{1.40}$$

The force constitutive equation $\mathbf{f}_{cyl}\left(\delta_n, \dot{\delta}_n, \delta_t, \dot{\delta}_t, \theta_{am}, \dot{\theta}_{am}\right)$ may involve the coupling of the relative displacements and misalignments. It also involves the clearance size, bearing and journal geometry and the material constitutive properties. For the complete definition of the cylindrical joint a penalization moment must be considered, written as

$$\begin{aligned} \mathbf{n}'^{(cyl)}_i &= \mathbf{n}_{cyl,i}\left(\delta_n, \dot{\delta}_n, \delta_t, \dot{\delta}_t, \theta_{am}, \dot{\theta}_{am}\right) \\ \mathbf{n}'^{(cyl)}_j &= \mathbf{n}_{cyl,j}\left(\delta_n, \dot{\delta}_n, \delta_t, \dot{\delta}_t, \theta_{am}, \dot{\theta}_{am}\right) \end{aligned} \tag{1.41}$$

A clearance/bushing, revolute joint requires the penalization forces and moments of the cylindrical joints and the penalization of the axial displacement as

$$\begin{aligned} \mathbf{f}^{(ad)}_i &= f_{ad}\left(\delta_t, \dot{\delta}_t\right)\mathbf{u}_{ad} \\ \mathbf{f}^{(ad)}_j &= -\mathbf{f}^{(ad)}_i \end{aligned} \tag{1.42}$$

The force relation $f_{ad}\left(\delta_t, \dot{\delta}_t\right)$ involves the axial displacement, its speed and the geometric and material characteristics of the joint.

The clearance/bushing translation joint besides the penalization forces and moments of the cylindrical joints also involves the penalization of the axial rotation, given by

$$\begin{aligned} \mathbf{n}'^{(ar)}_i &= f_{ar}\left(\alpha, \dot{\alpha}\right)\mathbf{A}_i^T\mathbf{b}_i \\ \mathbf{n}'^{(ar)}_j &= -\mathbf{A}_j^T\mathbf{A}_i\mathbf{n}'^{(ar)}_i \end{aligned} \tag{1.43}$$

where $f_{ar}\left(\alpha, \dot{\alpha}\right)$ is a nonlinear relation involving the axial rotation angle, its speed and the geometric and material characteristics of the joint.

The forces and moments of the clearance/bushing constitutive equations are applied in the bodies connected by the joint, i.e., on points P_i and P_j, in bodies i and j, respectively. The contribution of the clearance/bushing joints to the force vector of the bodies connected by one of those joints is

$$\begin{aligned} \mathbf{g}_i &= \left\{ \begin{array}{c} \left(\mathbf{f}^{(cyl)}_i + \mathbf{f}^{(ad)}_i\right) \\ \mathbf{s}'^P_i\mathbf{A}_i^T\left(\mathbf{f}^{(cyl)}_i + \mathbf{f}^{(ad)}_i\right) + \mathbf{n}'^{(cyl)}_i + \mathbf{n}'^{(ar)}_i \end{array} \right\}; \\ \mathbf{g}_j &= \left\{ \begin{array}{c} \mathbf{f}^{(cyl)}_j + \mathbf{f}^{(ad)}_j \\ \mathbf{s}'^P_j\mathbf{A}_j^T\left(\mathbf{f}^{(cyl)}_j + \mathbf{f}^{(ad)}_j\right) + \mathbf{n}'^{(cyl)}_j + \mathbf{n}'^{(ar)}_j \end{array} \right\} \end{aligned} \tag{1.44}$$

All forces and moments defined in Eqs. (1.40, 1.41, 1.42 and 1.43) involve the relative displacements and rotations and their time derivatives, for the evaluation of which the reader is directed to reference [48]. In the definition of the penalty moments, defined by Eqs. (1.41) and (1.43) the penalization of the axial displacement and axial rotation are decoupled from each other, and from the other relative

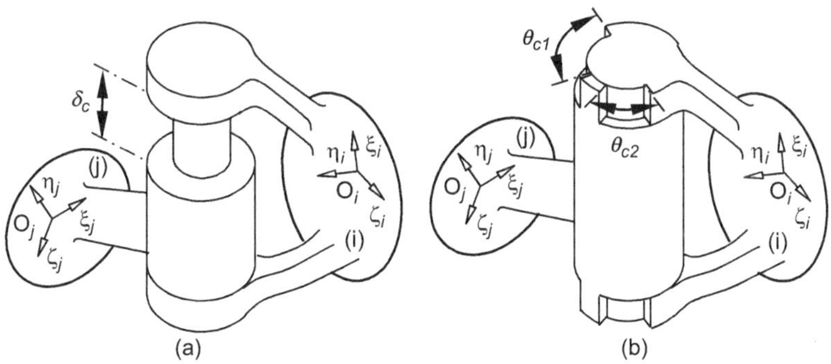

Fig. 1.24 Kinematic joints with limits in their range of motion: (**a**) cylindrical joint with limits on its translation displacement; (**b**) revolute joint with limits on its rotation range

motions. Depending on the material characteristics of the connected bodies, as in some biological applications or when involving some types of composite materials, some particular applications, the relative motion components may be coupled.

1.4.2 Joint Motion Limits

In practical applications the translation, cylindrical and revolute joints have limits on the relative motion between the two bodies, such as those illustrated in Fig. 1.24. The definition of the joint motion limits, for perfect kinematic joints or for clearance/bushing joints, are represented as a penalization of the joint motion after a given allowed range. From this point of view there is no difference in the methodology for the motion limit modeling in perfect kinematic or clearance/bushing joints. The penalization due to the joint stops is introduced in the force vector of the bodies that share the joint one of the relations defined by Eqs. (1.40, 1.41, 1.42 and 1.43), but considering the clearance size as the allowed range of motion.

For a perfect kinematic translation joint, with limits in the translation δ_c shown in Fig. 1.24a, the contributions to the system constraint equations and force vector

$$\begin{array}{ll}\text{constraint equations}: & \mathbf{\Phi}^{(trans,5)} = \mathbf{0} \\ \text{force vector}: & \begin{cases} \mathbf{f}_i = \mathbf{f}_i^{(ad)}; \;\; \mathbf{f}_j = -\mathbf{f}_i^{(ad)} \\ \mathbf{n}'_i = \mathbf{A}_i^T \tilde{\mathbf{s}}_i^P \mathbf{f}_i; \;\; \mathbf{n}'_j = \mathbf{A}_j^T \tilde{\mathbf{s}}_j^P \mathbf{f}_j \end{cases}\end{array} \tag{1.45}$$

while for a clearance/bushing translation joint the kinematic constraint is substituted by the forces defined by Eqs. (1.40, 1.41, 1.42 and 1.43), applied as implied in Eq. (1.44).

For a perfect revolute joint with limits on its rotation given by $\theta_c = \theta_{c1}+\theta_{c2}$, as shown in Fig. 1.24b, the contributions for the constraint equations and force vector are

$$\begin{aligned} &\text{constraint equations}: \boldsymbol{\Phi}^{(rev,5)} = \mathbf{0} \\ &\text{force vector}: \quad \begin{cases} \mathbf{f}_i = \mathbf{0};\ \mathbf{f}_j = \mathbf{0} \\ \mathbf{n}'_i = \mathbf{n}'^{(ar)}_i;\ \mathbf{n}'_j = -\mathbf{A}_j^T \mathbf{A}_i \mathbf{n}'^{(ar)}_i \end{cases} \end{aligned} \tag{1.46}$$

For a clearance/bushing revolute joint the kinematic constraint, appearing in Eq. (1.46) is eliminated and, instead, the penalization forces characteristics of the joint, are used.

It should be noted that when the joint motion limits are hard stops the penalization of the relative motion is defined without involving the bushing in the penalization force constitutive relation. For joint limits defined with soft stops, such as the bounce stops in the vehicle suspensions, the penalization force model must take into account both bushing and hard contact, defined according to their physical characteristics.

1.4.3 Application to a 3D Slider-Crank Mechanism

The spatial slider-crank mechanism presented in Fig. 1.25 is part of the library of multibody benchmark problems [93]. The detailed slider-crank model and the initial conditions for the simulation are described in reference [50]. To demonstrate the use of clearance/bushing joints in the modeling of multibody systems several models are built either considering all its joints as perfect or including an imperfect revolute joint between ground and crank, being the clearance varied. The joints with a null clearances are still modeled as clearance/bushing joints, although it is expected that they still behave as perfect revolute joints, as those modelled with kinematic constraints.

The normal contact force model with hysteresis damping, expressed by Eq. (1.26), is used in the definition of the force penalization of the revolute joint.

Fig. 1.25 Spatial slider-crank mechanism

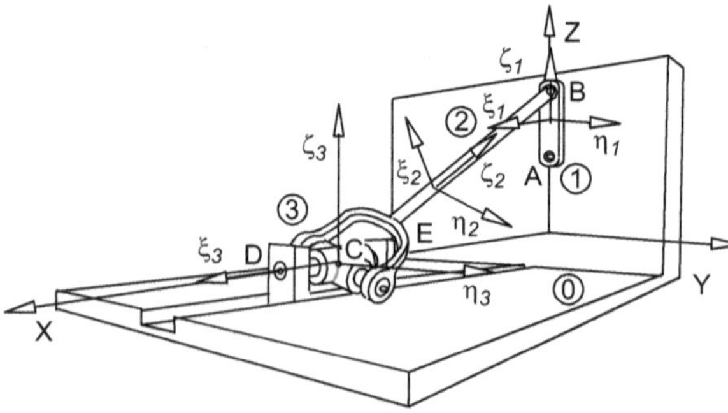

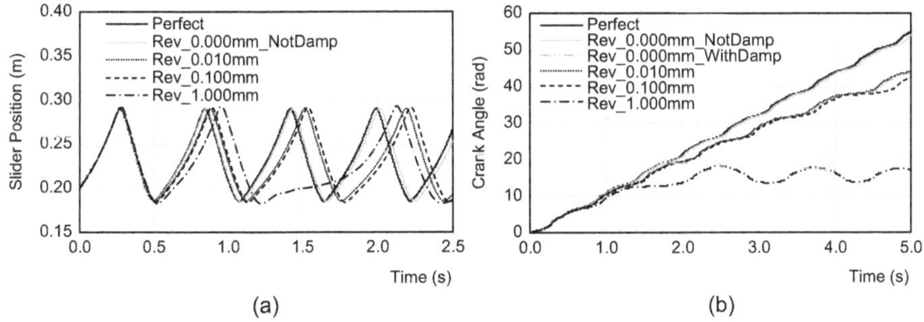

Fig. 1.26 Slider-crank model with revolute clearance joint: (**a**) slider position; (**b**) crank angle

The MUltiBOdy Dynamic analysis program MUBODyn [94] is used to develop the dynamic analyses of the models developed. The equations of motion of the system, together with the acceleration constraint equations, making a system of linear equations, are first solved for the unknown accelerations and Lagrange multipliers with a sparse matrix solver [96]. The velocities and accelerations are integrated numerically with the Gear multistep integrator [95]. The time stepping physical control procedure [69] is applied to detect the start of contact in all dynamic analyses for 5 s of simulation time.

The time histories of the slider block position and crank angle of the models are presented in Fig. 1.26. The results show that as the clearance in the revolute joint starts to increase, the time at which the slider reaches its end of stroke is increasingly delayed. For a clearance of 1 mm it is observed, in Fig. 1.26b, that the crank angle is unable to develop more than two revolutions, after which the crank ends up oscillating about its static equilibrium position. The effect of the energy dissipation due to the inclusion of the non-elastic restitution parameter in the contact force model is visible when comparing the responses of the models with a null clearance revolute joint with and without damping. The existence of the damping is responsible, just by itself, for the energy dissipation that forces the slider oscillation to increase the delay observed on reaching its end of stroke.

The joint reaction forces in the revolute joint are shown in Fig. 1.27. For the models with perfect revolute joints, modeled either as revolute kinematic constraints or as imperfect joints with null clearance, no oscillations in the reaction force are observed. As the clearance increases the joint reaction force of the revolute joint exhibits oscillations, mostly during the slider mid-stroke, being the amplitude of the force oscillations higher with the clearance size.

The presence of hysteresis damping in the normal contact model of the clearance joint eliminates the oscillations of the joint reaction forces. It has been observed in some cases that the friction forces, when present, also mitigate the oscillatory behavior observed in the contact joint [97]. The use of different numerical integrator schemes, with fixed or variable time steps or with internal damping can also mask, or emphasize, the oscillatory behavior observed.

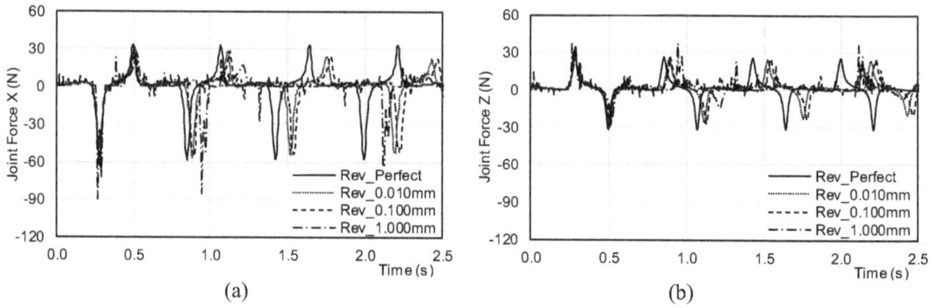

Fig. 1.27 Joint reaction force for the revolute joint slider crank models with perfect and with clearance revolute joint: (**a**) force component X; (**b**) force component Z

1.5 Highlights and Conclusions

This work identifies and discusses selected challenges on the formulation and application of multibody dynamics for important engineering systems. Important challenges, such as the computational geometry in kinematics and contact, formulation of complex kinematic joints, contact and impact in multibody dynamics, realistic modelling of common mechanical joints that exhibit clearances and local deformations, numerical issues associated with time integration or with nonlinear equations with physical inspired solution control, are just some of the topics which require a continued investigation for more advanced multibody systems. This work emphasized the importance of accurate and computational efficient developments in computational geometry to support other challenges in the modelling of multibody systems. The parametrization of curves and surfaces and their continuity not only need to fullfil the criteria associated to the specific application but also limit the numerical methods that can be applied in their solution. It has been observed that when the spatial curves, and surfaces, are used to setup kinematic constraints the minimum degree required for the polynomial description of the curve is that associated to the highest parametric derivative. However, when the same curve or surface is used in the context of contact mechanics the minimum continuity required depends on the solution method used for the contact detection. It has been shown that the use of nonlinear system of equations solvers, such as those implemented in the Matlab functions `fsolve` or `fmincon` only the parametric description of the curves or surfaces and their moving frame vectors, but are computationally more expensive than using the Newton-Raphson method. The drawback is that the Newton-Raphson method requires the derivatives of the parametric surfaces and associated vectors, putting higher demands on the degree of the parametric descriptions used. In any case, for large surfaces or long curves, eventually with a very fine discretization in terms of control nodes, it is not always possible, or efficient from the computational point of view, to keep them in memory. The use of look-up-tables, which allow that the contact search and the contact force evaluation to be done by linear interpolation of predefined nodes, avoid that complex operations need to be carried online. The development of clearance/bushing joints in which the mechanical tolerances used in the construction of machines and mechanisms are reflected allow for the development of realistic models of multibody systems that would not be validated otherwise. The general formulation of this joints allows for the introduction of local deformation such as that exhibited by bushing joints. In all implementations of contact detection problems the numerical issues associated with time integration play a major role. It has been shown that adjusting the time-step of the time integrators, in particular those that use a variable time-step, is of fundamental importance. It is reinforced here the proposal that a physical control of the time-step, to prevent the initialization of a contact with a large interference between contacting surfaces, must be available complementary to the common mathematical control internal to these algorithms.

References

1. Hooker, W., Margulies, G.: The dynamical attitude equations for n-body satellite. J. Astronaut. Sci. **12**, 123–128 (1965)
2. Kane, T., Scher, M.: A dynamical explanation of the falling cat phenomenon. Int. J. Solids Struct. **5**(7), 663–670 (1969)
3. Kane, T., Scher, M.: Human self-rotation by means of limb movements. J. Biomech. **3**(1), 39–49 (1970)
4. Wittenburg, J.: The dynamics of systems of coupled rigid bodies. A new general formalism with applications. In: Grioli, G. (ed.) Stereodynamics. Edizione Cremonese, Roma (1971)
5. Wittenburg, J., Wolz, U., Schmidt, A.: MESA VERDE – a general-purpose program package for symbolical dynamics simulations of multibody systems. In: Schiehlen, W. (ed.) Multibody Systems Handbook. Springer, Heidelberg (1990)
6. Magnus, K.: Dynamics of Multibody Systems. Proceedings of the IUTAM Symposium, Munich, Germany, August 29 – September 3, 1977, Springer, Heidelberg (1978)
7. Haug, E.J. (ed.): Computer Aided Analysis and Optimization of Mechanical Systems Dynamics. Springer-Verlag, Heidelberg (1984)
8. Bianchi, G., Schiehlen, W. (eds.): Dynamics of Multibody Systems, Proceedings of the IUTAM/IFToMM Symposium, Udine, Italy, September 16–20, 1985. Springer, Heidelberg (1986)
9. Pereira, M., Ambrosio, J. (eds.): Computer-Aided Analysis of Rigid and Flexible Mechanical Systems, NATO Science Series E, vol. 268. Springer, Dordrecht (1994)
10. Schiehlen, W.: Multibody dynamics: roots and perspectives. Multi-body Syst. Dyn. **1**(2), 149–188 (1997)
11. Wittenburg, J.: Dynamics of Systems of Rigid Bodies. Teubner-Verlag, Wiesbaden (1977)
12. Nikravesh, P.E.: Computer-Aided Analysis of Mechanical Systems. Prentice-Hall, Englewood Cliffs (1988)
13. Kane, T., Levinson, D.: Dynamics: Theory and Applications. McGraw-Hill, San Francisco (1985)
14. Haug, E.: Computer Aided Kinematics and Dynamics of Mechanical Systems. Allyn and Bacon, Boston (1989)
15. Jalon, G., Bayo, E.: Kinematic and Dynamic Simulation of Multibody Systems: The Real-Time Challenge. Springer, Berlin (1994)
16. Erdman, A.G., Sandor, G.N.: Kineto-elastodynamics – a review of the state of the art and trends. Mech. Mach. Theory. **7**, 19–33 (1972)

17. Lowen, G.G., Chassapis, C.: Elastic behavior of linkages: an update. Mech. Mach. Theory. **21**, 33–42 (1986)
18. Thompson, B.S., Sung, G.N.: Survey of finite element techniques for mechanism design. Mech. Mach. Theory. **21**, 351–359 (1986)
19. Song, J.O., Haug, E.J.: Dynamic analysis of planar flexible mechanisms. Comput. Methods Appl. Mech. Eng. **24**, 359–381 (1980)
20. Shabana, A.: Dynamics of Multibody Systems. Wiley, New York (1989)
21. Shabana, A., Wehage, R.: A coordinate reduction technique for transient analysis of spatial structures with large angular rotations. J. Struct. Mech. **11**, 401–431 (1989)
22. Meirovitch, L., Nelson, H.D.: On the high-spin motion of a satellite containing elastic parts. J. Spacecr. Rocket. **3**, 1597–1602 (1966)
23. Modi, V., Suleman, A., Ng, A.: An approach to dynamics and control of orbiting flexible structures. Int. J. Numer. Methods Eng. **32**, 1727–1748 (1991)
24. Banerjee, A.K., Nagarajan, S.: Efficient simulation of large overall motion of nonlinearly elastic beams. In: Proceedings of ESA International Workshop on Advanced Mathematical Methods in the Dynamics of Flexible Bodies, ESA, Noordwijk, The Netherlands (1996)
25. Kane, T., Ryan, R., Banerjee, A.: Dynamics of a cantilever beam attached to a moving base. AIAA J. Guid. Control Dyn. **10**, 139–151 (1987)
26. Wallrapp, O., Schwertassek, R.: Representation of geometric stiffening in multibody system simulation. Int. J. Numer. Methods Eng. **32**, 1833–1850 (1991)
27. Geradin, M.: Advanced methods in flexible multibody dynamics: review of element formulations and reduction methods. In: Proceedings of ESA International Workshop on Advanced Mathematical Methods in the Dynamics of Flexible Bodies, ESA, Noordwijk, The Netherlands (1996)
28. Belytschko, T., Hsieh, B.J.: Nonlinear transient finite element analysis with convected coordinates. Int. J. Numer. Methods Eng. **7**, 255–271 (1973)
29. Simo, J.C., Vu-Quoc, L.: On the dynamics in space of rods undergoing large motions – a geometrically exact approach. Comp. Methods Appl. Mech. Eng. **66**, 125–161 (1988)
30. Bathe, K.-J., Bolourchi, S.: Large displacement analysis of three-dimensional beam structures. Int. J. Numer. Methods Eng. **14**, 961–986 (1979)
31. Cardona, A., Geradin, M.: A beam finite element non linear theory with finite rotations. Int. J. Numer. Methods Eng. **26**, 2403–2438 (1988)
32. Geradin, M., Cardona, A.: A modelling of superelements in mechanism analysis. Int. J. Numer. Methods Eng. **32**, 1565–1594 (1991)
33. Shabana, A.: Definition of the slopes and the finite element absolute nodal coordinate formulation. Multi-body Syst. Dyn. **1**, 339–348 (1997)
34. Ambrósio, J., Nikravesh, P.: Elastic-plastic deformations in multibody dynamics. Nonlinear Dyn. **3**, 85–104 (1992)
35. Ambrósio, J.: Dynamics of structures undergoing gross motion and nonlinear deformations: a multibody approach. Comput. Struct. **59**(6), 1001–1012 (1996)
36. Pereira, M.S., Ambrósio, J.: Crashworthiness analysis and design using rigid-flexible multibody dynamics with application to train vehicles. Int. J. Numer. Methods Eng. **40**(4), 655–687 (1997)
37. Shabana, A.: Flexible multibody dynamics: review of past and recent developments. Multibody Syst. Dyn. **1**(2), 189–222 (1997)
38. Geradin, M., Cardona, E.: Flexible Multibody Dynamics: A Finite Element Approach. Wiley, Chichester (2001)
39. Bauchau, O.: Flexible Multibody Dynamics. Springer, Dordrecht (2011)
40. Bremer, H.: Elastic Multibody Dynamics: A Direct Ritz Approach. Springer, Dordrecht (2008)
41. Gonzalez-Palacios, M., Angeles, J.: Cam Synthesis. Springer, Dordrecht (1993)
42. Gonzalez-Palacios, M., Angeles, J.: Synthesis of contact surfaces of spherical cam-oscillating roller-follower mechanisms. ASME J. Mech. Des. **116**(1), 315–319 (1994)
43. Pombo, J., Ambrósio, J.: General spatial curve joint for rail guided vehicles: kinematics and dynamics. Multi-body. Syst. Dyn. **9**(3), 237–264 (2003)

44. Ambrósio, J., Antunes, P., Pombo, J.: On the requirements of interpolating polynomials for path motion constraints. In: Kecskeméthy, A., Geu Flores, F. (eds.) Interdisciplinary Applications of Kinematics: Proceedings of the International Conference, pp. 179–197. Springer, Dordrecht (2015)
45. Tändl, M., Kecskemethy, A.: Singularity-free trajectory tracking with Frenet frames. In: Husty, M., Schroecker, H.-P. (eds.) Proceedings of the 1st Conference EuCoMeS. Innsbruck University Press, Obergurgl (2006)
46. Tändl, M.: Dynamic Simulation and Design of Roller Coaster Motion. VDI Verlag, Düsseldorf (2009)
47. Flores, P., Ambrósio, J., Pimenta Claro, J., Lankarani, H.: Kinematics and Dynamics of Multibody Systems with Imperfect Joints. Springer, Dordrecht (2008)
48. Ambrósio, J., Verissimo, P.: Improved bushing models for vehicle dynamics. Multi-body Syst. Dyn. **22**(4), 341–365 (2009)
49. Magalhaes, H., Ambrósio, J., Pombo, J.: Railway vehicle modelling for the vehicle-track interaction compatibility analysis. Proc. Inst. Mech. Eng. K J. Multi-body Dyn. **230**(3), 251–267 (2016)
50. Ambrósio, J., Pombo, J.: A unified formulation for mechanical joints with and without clearances/bushings and/or stops in the framework of multibody systems. Multi-body Syst. Dyn. **42**(3), 317–345 (2018)
51. Akhadkar, N., Acary, V., Brogliato, B.: Multibody systems with 3D revolute joints with clearances: an industrial case study with an experimental validation. Multi-body Syst. Dyn. **42**(3), 249–282 (2018)
52. Ambrósio, J.: Efficient kinematic joints descriptions for flexible multibody systems experiencing linear and non-linear deformations. Int. J. Numer. Methods Eng. **56**, 1771–1793 (2003)
53. Masarati, P., Morandini, M.: Intrinsic deformable joints. Multi-body Syst. Dyn. **23**, 361–386 (2010)
54. Cardona, A., Geradin, M., Doan, D.B.: Rigid and flexible joint modelling in multibody dynamics using finite elements. Comput. Methods Appl. Mech. Eng. **89**(1–3), 395–418 (1991)
55. Bae, D., Han, J., Choi, J.: An implementation method for constrained flexible multibody dynamics using virtual body and joint. Multi-body Syst. Dyn. **4**, 207–226 (2000)
56. Gonçalves, J., Ambrósio, J.: Advanced modeling of flexible multibody dynamics using virtual bodies. Comput. Assist. Mech. Eng. Sci. **9**(3), 373–390 (2002)
57. Mashayekhi, M., Kövecses, J.: A comparative study between the augmented Lagrangian method and the complementarity approach for modeling the contact problem. Multi-body Syst. Dyn. **40**(4), 327–345 (2017)
58. Blumentals, A., Brogliato, B., Bertails-Descoubes, F.: The contact problem in Lagrangian systems subject to bilateral and unilateral constraints, with or without sliding Coulomb's friction: a tutorial. Multi-body Syst. Dyn. **38**(1), 43–76 (2016)
59. Zhao, Z., Liu, C.: Contact constraints and dynamical equations in Lagrangian systems. Multi-body Syst. Dyn. **38**(1), 77–99 (2016)
60. Flores, P., Leine, R., Glocker, C.: Modeling and analysis of planar rigid multibody systems with translational clearance joints based on the non-smooth dynamics approach. Multi-body Syst. Dyn. **23**(2), 165–190 (2010)
61. Kwak, S.D., Blankevoort, L., Ateshian, G.A.: A mathematical formulation for 3D quasi-static multibody models of diarthrodial joints. Comput. Methods Biomech. Biomed. Eng. **3**, 41–64 (2000)
62. Bei, Y., Fregly, B.J.: Multibody dynamic simulation of knee contact mechanics. Med. Eng. Phys. **26**(9), 777–789 (2004)
63. Machado, M., Flores, P., Pimenta Claro, J.C., Ambrósio, J., Silva, M., Completo, A., Lankarani, H.: Development of a planar multibody model of the human knee joint. Nonlinear Dyn. **60**(3), 459–478 (2010)
64. Anand, V.: Computer Graphics and Geometric Modeling for Engineers. Wiley, New York (1996)
65. Frenet, F.: Sur les courbes à double courbure. J. Math. Pures Appl. **17**, 437–447 (1852)

66. Machado, M., Flores, P., Ambrósio, J.: A lookup table-based approach for spatial analysis of contact problems. J. Comput. Nonlinear Dyn. **9**(1), 1–10 (2014)
67. Viegas, M., Ambrósio, J., Antunes, P., Magalhães, H.: Dynamics of a roller coaster vehicle. In: Spriyagin, M., Gordon, T., Cole, C., McSweeney, T. (eds.) Proceedings of the 25th International Symposium on Dynamics of Vehicles on Roads and Tracks (IAVSD 2017), Volume 2, pp. 551–556. CRC Press, Taylor and Francis, London (2017)
68. Kikuuwe, R., Brogliato, B.: A new representation of systems with frictional unilateral constraints and its Baumgarte-like relaxation. Multi-body Syst. Dyn. **39**(3), 267–290 (2017)
69. Flores, P., Ambrósio, J.: On the contact detection for contact-impact analysis in multibody systems. Multi-body Syst. Dyn. **24**(1), 103–122 (2010)
70. Haddouni, M., Acary, V., Garreau, S., Beley, J.-D., Brogliato, B.: Comparison of several formulations and integration methods for the resolution of DAEs formulations in event-driven simulation of nonsmooth frictionless multibody dynamics. Multi-body Syst. Dyn. **41**(3), 201–231 (2017)
71. Pombo, J., Ambrósio, J.: Application of a wheel–rail contact model to railway dynamics in small radius curved tracks. Multi-body Syst. Dyn. **19**(1), 91–114 (2008)
72. Pacejka, H.: Tyre and Vehicle Dynamics, 3rd edn. Butterworth-Heinemann, Amsterdam (2012)
73. Hirschberg, W., Rill, G., Weinfurter, H.: Tire model TMeasy. Veh. Syst. Dyn. **45**(1), 101–119 (2007)
74. Kalker, J.J.: Three-Dimensional Elastic Bodies in Rolling Contact. Springer, Dordrecht (1990)
75. Harris, T., Kotzalas, M.: Advanced Concepts of Bearing Technology. CRC Press, Boca Raton (2007)
76. Lankarani, H., Nikravesh, P.: Continuous contact force models for impact analysis in multibody systems. Nonlinear Dyn. **5**(2), 193–207 (1994)
77. Hunt, K.H., Grossley, F.R.E.: Coefficient of restitution interpreted as damping in vibroimpact. ASME J. Appl. Mech. **7**, 440–445 (1975)
78. Ambrósio, J.: Rigid and flexible multibody dynamics tools for the simulation of systems subjected to contact and impact conditions. Eur. J. Solids A Solids. **19**(S), 23–44 (2000)
79. Marques, F., Flores, P., Pimenta Claro, J., Lankarani, H.: A survey of several friction force models for dynamic analysis of multibody mechanical systems. Nonlinear Dyn. **86**(3), 1407–1443 (2016)
80. Bengisu, M.T., Akay, A.: Stability of friction-induced vibrations in multi-degree-of-freedom systems. J. Sound Vib. **171**, 557–570 (1994)
81. Kelley, C.T.: Iterative Methods for Linear and Nonlinear Equations. SIAM, Philadelphia (1995)
82. Gupta, P.K.: Advanced Dynamics of Rolling Elements. Springer, New York (1984)
83. Haines, D., Ollerton, E.: Contact stress distribution on elliptical contact surfaces subjected to radial and tangential forces. Proc. Inst. Mech. Eng. **177**, 95–114 (1963)
84. Magalhães, H., Madeira, J., Ambrósio, J., Pombo, J.: Railway vehicle performance optimization using virtual homologation. Veh. Syst. Dyn. **54**(9), 1177–1207 (2016)
85. Polach, O.: A fast wheel-rail forces calculation computer code. Veh. Syst Dyn. **33**, 782–739 (1999)
86. Ayasse, J., Chollet, H.: Determination of the wheel rail contact patch in semi-Hertzian conditions. Veh. Syst. Dyn. **43**, 161–172 (2005)
87. Sichani, M.S., Enblom, R., Berg, M.: A novel method to model wheel-rail normal contact in vehicle dynamics simulation. Veh. Syst. Dyn. **52**, 1752–1764 (2014)
88. Piotrowski, J., Liu, B., Bruni, S.: The Kalker book of tables for non-Hertzian contact of wheel and rail. Veh. Syst. Dyn. **55**(6), 875–901 (2017)
89. Escalona, J., Aceituno, J.: Modeling wheel-rail contact with pre-calculated lookup tables in arbitrary-geometry tracks with irregularities. In: ASME Proceedings of the 11th International Conference on Multibody Systems, Nonlinear Dynamics, and Control, Boston, Massachusetts, USA, August 2–5, Paper No. DETC2015-47306 (2015)
90. Zhang, J., Wang, Q.: Modeling and simulation of a frictional translational joint with a flexible slider and clearance. Multi-body Syst. Dyn. **38**(4), 367–389 (2016)

91. Pichler, F., Witteveen, W., Fischer, P.: A complete strategy for efficient and accurate multibody dynamics of flexible structures with large lap joints considering contact and friction. Multibody Syst. Dyn. **40**(4), 407–436 (2017)
92. Wang, G., Qi, Z., Wang, J.: A differential approach for modeling revolute clearance joints in planar rigid. Multi-body Syst. Dyn. **39**(4), 311–335 (2017)
93. Masoudi, R., Uchida, T., Vilela, D., Luaces, A., Cuadrado, J., McPhee, J.: A library of computational benchmark problems for the multibody dynamics community. In: Terze, Z. (ed.) Proceedings of ECCOMAS Multibody Dynamics., 1–4 July, pp. 1153–1162. University of Zagreb, Croatia (2013)
94. Ambrósio, J., Pombo, J.: MUltiBOdy Dynamic Analysis Program – MUBODyn: User's Manual, Technical Report IDMEC-CPM. Instituto de Engenharia Mecânica, Instituto Superior Técnico, University of Lisbon, Lisbon (2016)
95. Gear, G.: Numerical simulation of differential-algebraic equations. IEEE Trans. Circ Theory. **18**, 89–95 (1981)
96. Duff, I., Erisman, A., Reid, J.: Direct Methods for Sparse Matrices. Clarendon Press, Oxford (1986)
97. Ambrósio, J., Malça, C., Ramalho, A.: Planar roller chain drive dynamics using a cylindrical contact force model. Mech. Based Des. Struct. Mach. **44**(1–2), 109–122 (2015)

Chapter 2
Musculo-skeletal Modeling and Analysis for Low-Cost Active Orthosis Customization and SCI Patient Adaptation

Javier Cuadrado, Urbano Lugris, Francisco Mouzo, and Florian Michaud

Abstract This paper describes the role played by skeletal and musculo-skeletal modeling and analysis in: (i) the customization of low-cost active knee-ankle-foot orthoses aimed at assisting the gait of spinal-cord-injured subjects possessing some level of hip actuation; (ii) the adaptation process of the patients to the devices. It is shown that personalized multibody-based human models and related methods and techniques are very helpful for motor-gearbox selection and swing motion definition, for monitoring progress during the training sessions, for evaluating the final outcome provided by the assistive devices, and for anticipating their long-term impact on the patient's health.

Keywords Exoskeleton · Spinal-cord injury · Human modeling and simulation

2.1 Introduction

It is estimated that around half a million individuals become spinal-cord injured (SCI) every year, due to different causes as road, sport and labour accidents, falls, violence, etc. Walking impairment after injury leads to a decreased quality of life and, usually, to several collateral health issues as chronic pain, vein thrombosis, urinary tract infections, pressure ulcers or respiratory complications, thus increasing the risk of premature death too. Moreover, there are important associated health care costs, both at individual and social levels. As essential measures to improve the quality of life of SCI subjects, the World Health Organization recommends to streamline their access to rehabilitation services and assistive devices so as to maximize function, independence, well-being and integration [1].

J. Cuadrado (✉) · U. Lugris · F. Mouzo · F. Michaud
Laboratory of Mechanical Engineering, University of La Coruña, Ferrol, Spain
e-mail: javier.cuadrado@udc.es; ulugris@udc.es; francisco.mouzo@udc.es; florian.michaud@udc.es

E. Zahariev, J. Cuadrado (eds.), *IUTAM Symposium on Intelligent Multibody Systems – Dynamics, Control, Simulation*, IUTAM Bookseries 33,
https://doi.org/10.1007/978-3-030-00527-6_2

Rehabilitation services are generally provided in health-care centers, either by specialized personnel or dedicated machines. Assistive devices which are economically affordable for national health systems or for patients, usually show a passive behavior and do not take advantage of recent advances on mechatronics. On the other hand, modern exoskeletons are too expensive to reach the average patient [2]. Therefore, a new generation of low-cost active orthotic devices is needed to translate the huge potential of current mechatronics into a substantial jump in the quality of life of SCI subjects.

There are several types of orthotic devices depending on the type of injury they are targeted to: patients with full paraplegia (no or very weak actuation at hip, knee and ankle) need hip-knee-ankle-foot orthoses (HKAFO); patients with no or very weak actuation at knee and ankle require knee-ankle-foot orthoses (KAFO); and patients with no or very weak actuation at ankle demand ankle-foot orthoses (AFO). The present work focuses on the second case, i.e. patients with some hip actuation but no or very weak actuation at knee and ankle levels. The conventional KAFO which is generally prescribed to such patients is a passive device that has two functions: (a) locking the knee to avoid its flexion under the patient's weight during stance and the subsequent fall of the patient; (b) limiting the foot plantarflexion during swing to avoid stumbling due to forefoot collision with the ground.

Some patients are able to walk wearing the described passive KAFO, usually with the help of crutches for lateral stability (Fig. 2.1), their gait being faster or slower depending on the specific injury, the rehabilitation period, and the general physical and mental state. However, since knees are kept extended at all times, even during the swing phase, gait becomes very uncomfortable as the hip must be raised for the swing, thus leading to a high energetic cost which makes fatigue appear quickly. The result is that many patients prefer to use the wheelchair, thus losing the benefits of walking for rehabilitation and for their general health state.

Semi-active orthoses, called stance-control knee-ankle-foot orthoses (SCKAFO), can be found in the market that attempt to alleviate the mentioned problem: they

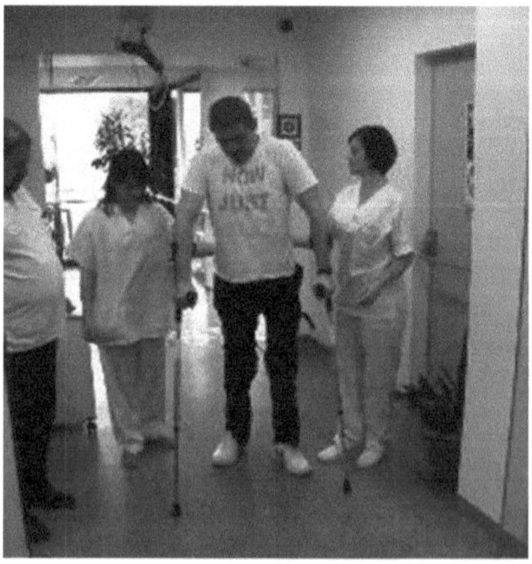

Fig. 2.1 Patient wearing passive KAFO at hospital

keep the knee locked during stance, but allow its flexion during gait, usually introducing some damping too for gait stabilization (as the biological knee does). Cost, bulkiness, weight, noise or requirement of patient's good pelvic control are some drawbacks of these devices [3]. Moreover, several proposals of active KAFO can be found in the literature [4, 5], which actuate the knee joint during swing. However, they are rather complex and, more importantly, as far as the authors know, not commercially available devices but just research prototypes.

2.2 The Low-Cost Active KAFO

The low-cost active KAFO used in this work is illustrated in Fig. 2.2. Starting from a conventional passive device, two actions are carried out to make it active: (a) an electric motor plus Harmonic gearbox are placed at knee level, substituting the external original joint; (b) an inertial measurement unit (IMU) is placed at shank level in the external upright. The total weight of each orthosis is kept under 2 kg. The orthoses are complemented by a backpack, carried by the patient, which contains an embedded computer board, the motor drivers and the battery. The weight of the backpack is also kept under 2 kg. Additionally, a smartphone application has been developed that allows to switch on/off the system, enable/disable the gait cycle, and setting the controller parameters.

The objective is that the system automatically detects the patient's swing motion intention so as to trigger the programmed swing cycle which must be executed by the motor. To this end, data from the IMU are received which, after going through a sensor fusion algorithm, provide the sagittal angular orientation and the vertical

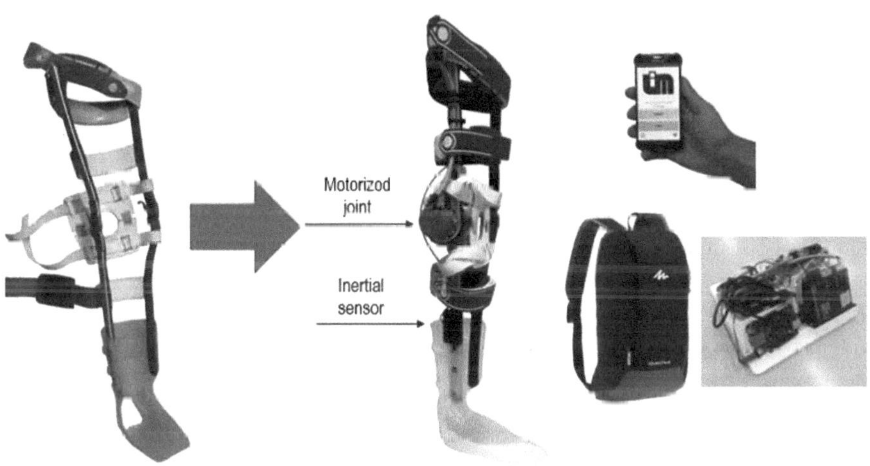

Fig. 2.2 The low-cost active KAFO

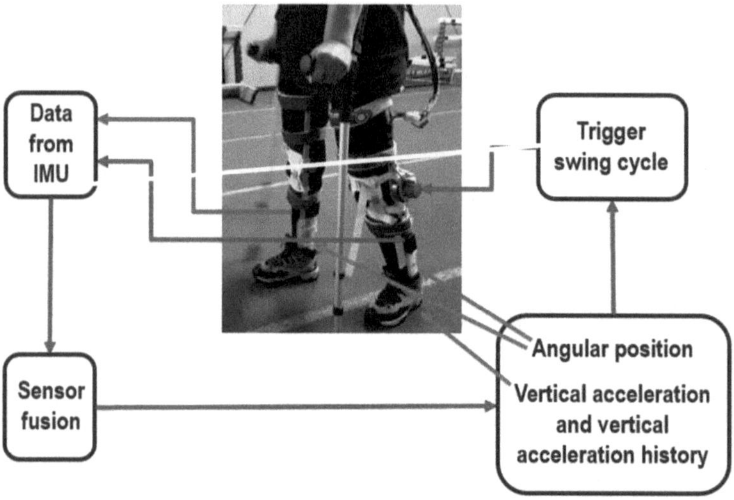

Fig. 2.3 The control scheme of the orthoses

acceleration of both shanks. The decision on whether or not to launch the swing cycle is taken upon the values of such magnitudes and their histories, adopting the necessary thresholds to avoid false positives which might cause the fall of the patient and are therefore unacceptable. This control scheme is illustrated in Fig. 2.3.

2.3 The Adaptation Process. Skeletal Models

Once the orthotic devices have been described, it is time to explain how they are personalized to the specific patient. To begin with, the selection criterion for the patient is that he/she possesses some hip actuation and no or almost no actuation at knee and ankle levels. As patients usually show partial hip actuation, they are asked to perform some hip flexion-extension exercises during which electromyography (EMG) is measured with surface electrodes in most possible muscles in order to determine which of them are active (Fig. 2.4).

Moreover, if the patient is able to walk with passive orthoses and crutches, a gait analysis is carried out in a lab which is prepared to capture such types of gaits. Particular features of the lab are the possibility of configuring the position of the force plates depending on the patient's gait characteristics, and the availability of a couple of crutches that have been instrumented in order to measure the ground reaction forces acting at their tips (Fig. 2.5).

A personalized computational skeletal model is developed which is animated with the captured motion (Fig. 2.6a). The model [6] (Fig. 2.6b) possesses 18 anatomical segments linked by ideal spherical joints, leading to 57 degrees of freedom: 3 pelvis translations plus 18 sets of 3 angles for the orientation of the segments. Crutches are rigidly connected to the hands while the orthoses are

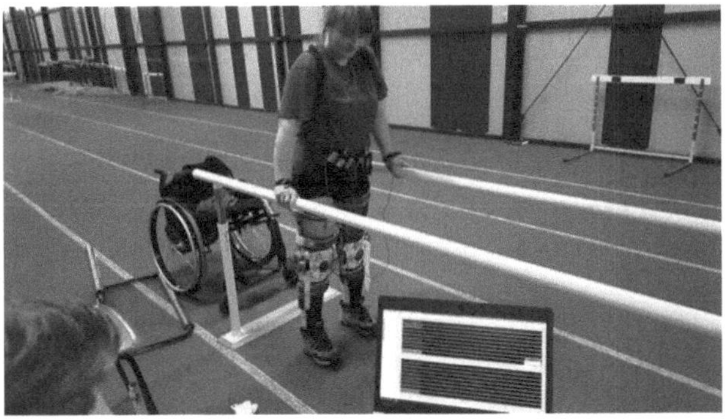

Fig. 2.4 Hip exercises to determine actuation level of hip muscles

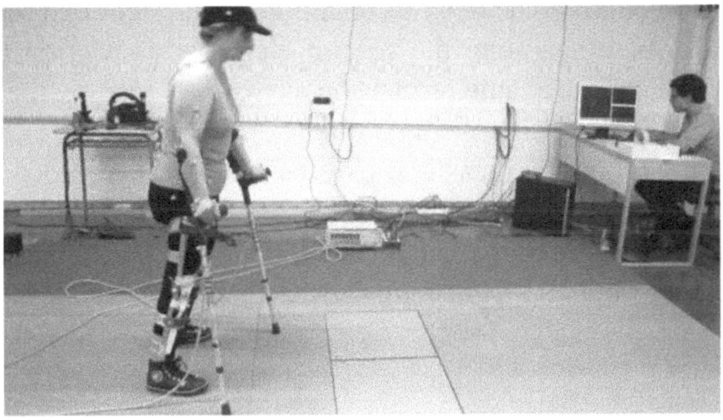

Fig. 2.5 Gait analysis with passive orthoses

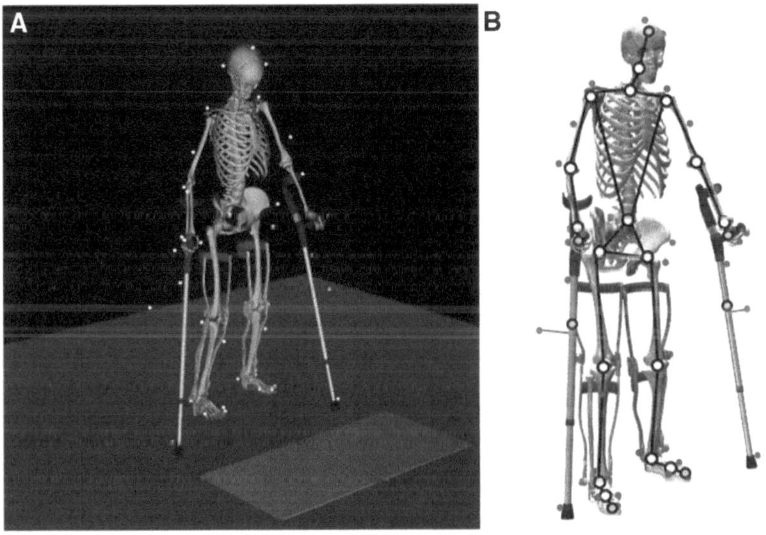

Fig. 2.6 Skeletal model: (**a**) animated with captured motion; (**b**) modeling details

embedded in the corresponding body links (thighs, shanks and feet). The inverse-dynamics analysis (IDA) of the model provides the histories of the joint drive torques and the external reactions. To this end, a velocity transformation formulation (called matrix-R [7]) is applied which yields the equations of motion in minimum number of coordinates.

In normal gait, IDA provides the ground reaction forces (GRF) at the supporting foot during the single support phase. However, when double support takes place, IDA only provides the total GRF, not being able to establish the GRF distribution between both feet. This effect is even more acute in the gait of patients using orthoses and crutches, since multi-contact between system and ground happens during most of the time (i.e. feet or crutches touching the ground). To overcome this problem, most authors [8–10] make use of the measured external forces, although this leads to the appearance of residuals, due to inconsistency among the model parameters, the captured motion and the measured external forces. The approach adopted here has been to obtain the consistent external reaction from IDA, and to distribute it among the several elements contacting the ground (in multi-support phases) in the same proportion as that provided by the measured external reactions. Furthermore, the obtained total external reaction from IDA can be compared with the sum of the measured external reactions in order to get a quality index of the performed analysis. See [6] for more detail about how IDA is carried out.

Results from the gait analysis are useful for system customization. Torques at the knees obtained from IDA (Fig. 2.7) indicate the torque required from the electric

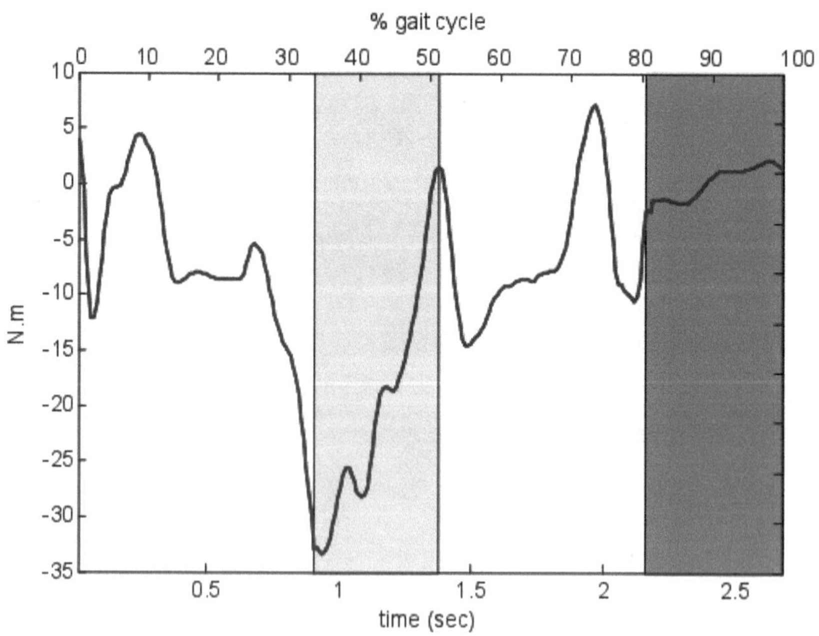

Fig. 2.7 History of sagittal torque in the patient's right knee. Grey areas indicate swing periods: right leg (dark); left leg (mild)

Fig. 2.8 Motor + gearbox and IMU placement. Training on parallel bars

motor and gearbox. The time employed by the patient in the swing, during which the motorized joint has the mission of flexing and extending the knee, yields the required speed. Combination of both magnitudes provides the required motor power and gearbox reduction. In this way, motor and gearbox can be selected. Moreover, the swing cycle (angle vs. time) is defined based on the obtained kinematics, and is sent to the motor controller for it to execute the motion (the motor controller adjusts its gains through the auto-tuning function, which is run with the patient wearing the orthoses before gait starts).

After taking measurements on the patient, the orthopedist builds the tailored uprights. The machined part containing the motor and gearbox is connected at knee level to the external thigh and shank uprights, and the IMU is fixed to the external shank upright (Fig. 2.8).

Then, the training sessions start. Initially, the patient walks with the help of parallel bars for the sake of safety (Fig. 2.8), and the swing cycle is progressively introduced in the devices as the patient gains confidence. Later, the same process is repeated without parallel bars, using crutches instead. In order to monitor the progress achieved by the patient along the training sessions, a set of parameters is selected. Examples of such parameters are gait speed, stride length, or lateral displacement of the center of mass. But many more are recorded. Since the training environment is a sport facility, not the gait analysis lab, they are obtained by means of a portable motion capture system (composed by six cameras) and subsequent post-processing by a computer-based application which helps the analyst in the task (Fig. 2.9).

Finally, after some training sessions, the patient is adapted to the use of the active orthoses (Fig. 2.10). The progress monitoring parameters can serve to compare the gait of the patient with the original passive orthoses and with their new active counterparts. Some representative parameters are gathered in Fig. 2.11 after being normalized for a more straightforward comparison.

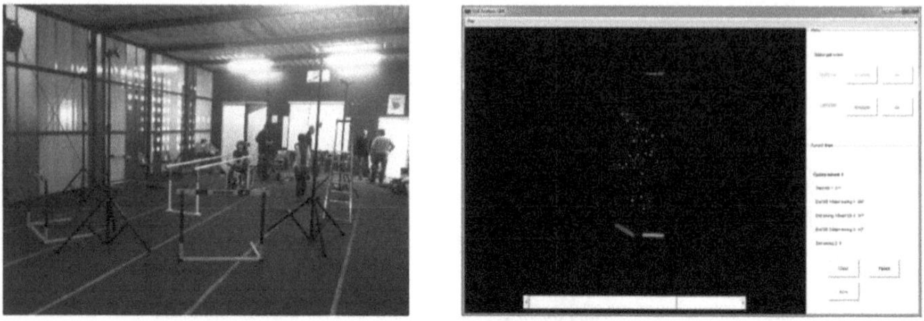

Fig. 2.9 Portable motion capture system and post-processing application

Fig. 2.10 Patient adapted to the use of the active orthoses

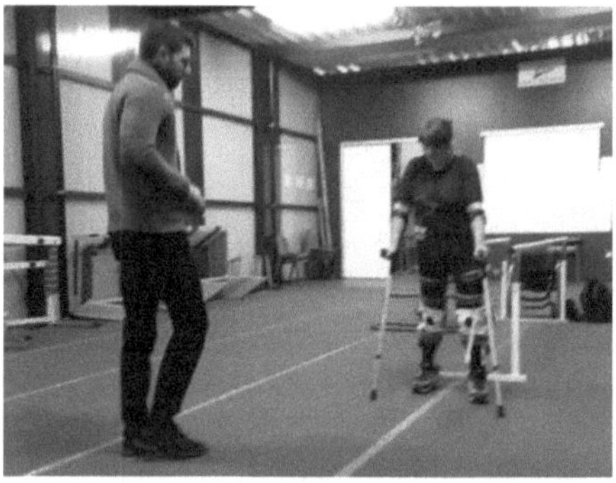

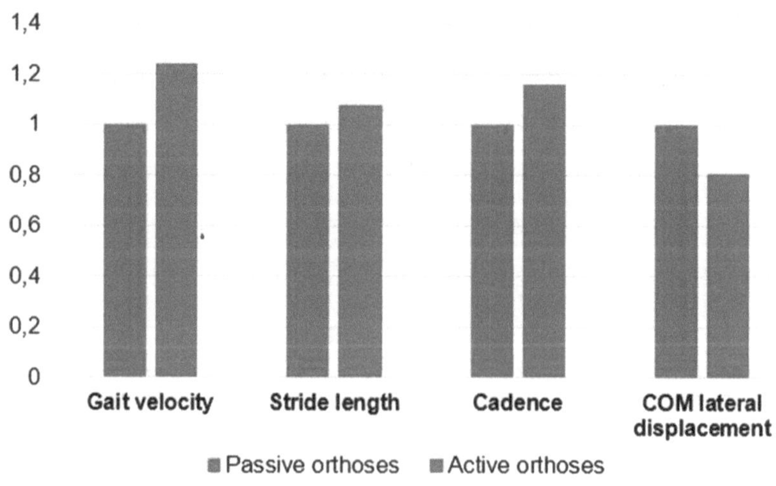

Fig. 2.11 Comparison of gaits: passive vs. active orthoses

2.4 Impact on the Patient's Health. Musculo-skeletal Models

Although customization of the active orthoses, along with the patient adaptation process, can be addressed with the help of just skeletal models as explained before, the use of musculo-skeletal models can provide information to evaluate the future impact that the active devices may have in patient's health. Including muscles in the model allows for the estimation of magnitudes as the joint reaction forces or the energetic cost. Joint reaction forces are important in these patients, especially in the upper limbs due to the use of crutches, which entails high loads during most part of the gait cycle. As for the energetic cost, its high value when wearing passive orthoses was said in the introduction to be one of the major causes of their rejection by patients.

The gait analysis process is illustrated in Fig. 2.12 for the case when muscles are taken into account in the model. In this case, EMG is also measured in as many muscles as possible during the tests, so as to provide a reference for checking the quality of the analysis.

Estimation of muscle forces is enough to determine joint reaction forces [11] (muscle excitation and activation are not needed) and, then, static optimization could be applied in order to solve the muscle recruitment problem. However, static optimization does not take into account muscle activation and contraction dynamics, being therefore preferable the so-called physiological static optimization which, besides fulfilling the previous condition, also provides the values of activations that may serve for comparison with measured EMG values and are also required by current energetic cost models. Physiological static optimization relies on the Hill's muscle model and, basically, attempts to establish dynamic consistency in

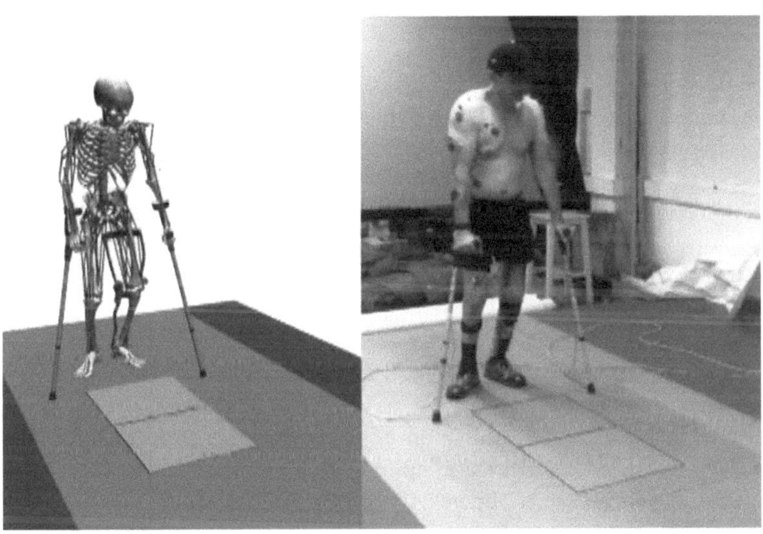

Fig. 2.12 Gait analysis with musculo-skeletal models

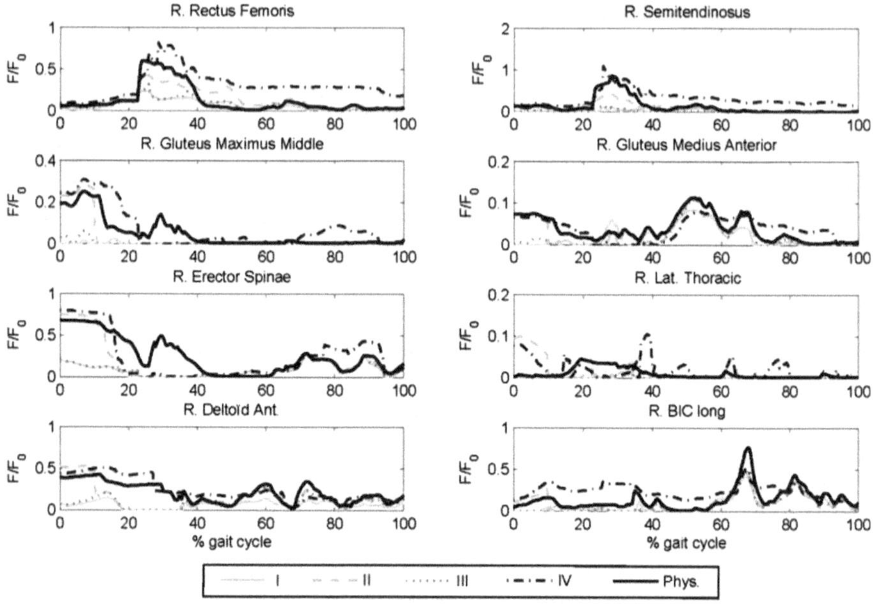

Fig. 2.13 Comparison of normalized muscle forces obtained by static (four criteria) and physiological static optimization (F: musculo-tendon force; F_0: maximum isometric force)

Table 2.1 The five muscle recruitment criteria

I	II	III	IV	Phys.
$\sum_{i=1}^{m} F_i^2$	$\sum_{i=1}^{m} \left(\frac{F_i}{F_{i,0}}\right)^2$	$\sum_{i=1}^{m} \left(\frac{F_i}{A_i}\right)^2$	$\max\left\{\frac{F_i}{F_{i,0}}\right\} \quad i = 1, 2, \ldots, m$	$\sum_{i=1}^{m} \left(\frac{F_i}{F_{i,\max}}\right)^2$

the history of muscle forces. To this end, it determines feasible force limits at each time-point by integrating activation and contraction equations along the time step for minimum and maximum values of the excitation. More detail about the static and physiological static optimization approaches performed here can be found in [12].

A comparison of normalized force histories for several muscles in the gait of the patient shown in Fig. 2.12 when estimated with static (four different criteria) and physiological static optimization is illustrated in Fig. 2.13. The corresponding cost functions are shown in Table 2.1, where F_i stands for the force in muscle i, $F_{i,0}$ is the maximum isometric force, A_i is the cross sectional area, and $F_{i,\max}$ is the maximum admissible force in muscle i at the particular time-point. As it can be observed, similar trends are obtained through the five methods. Furthermore, Fig. 2.14 shows the correlation between the estimation of the activations (for the same muscles) provided by the physiological static optimization and the measured EMG values, which further confirms the confidence in the obtained results.

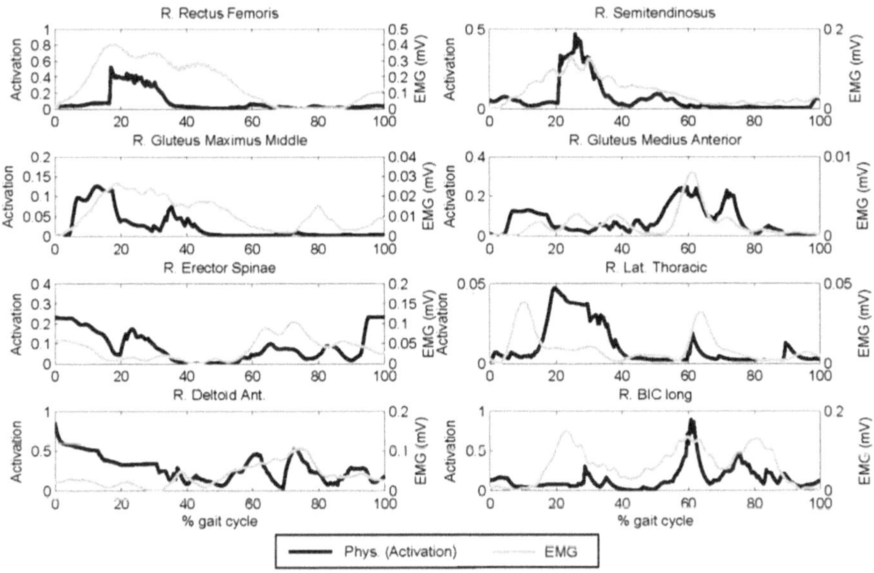

Fig. 2.14 EMG vs. muscular activations estimated by physiological static optimization

Making use of the estimated muscle forces, joint reactions can be obtained. Shoulders are probably the most demanded joint of the upper extremities when walking with crutches and, hence, they are prone to suffering injuries. The joint reaction history at the left shoulder of the patient shown in Fig. 2.12 is illustrated in Fig. 2.15, along with the activation history of one of the shoulder muscles. It can be seen that the load born by the shoulder is high during the full gait cycle, even surpassing the body weight at some instants. Moreover, both joint and muscles have very short rest phases.

The estimated muscle forces and activations can also feed models for energy consumption [13]. Tests with a portable gas analyzer were run to validate the output of such models in healthy gait, so that they can be later applied for the gait of injured subjects. Basically, the test consists of walking for 5 min in a circuit made with cones. A sound is generated every time the subject must pass by a cone, so that he/she can adjust the speed to the desired one. Figure 2.16 shows a test of a healthy subject wearing a gas analyzer, along with a comparison between results obtained from such experiments and results obtained by application of the energy expenditure model proposed in [13]. A constant offset was observed between the results provided by both methods, which could hence be calibrated. Moreover, some variability was found in the results based on the energy consumption model for a certain gait speed and subject, which made advisable to repeat the motion capture several times and average the corresponding energetic results. With such procedures,

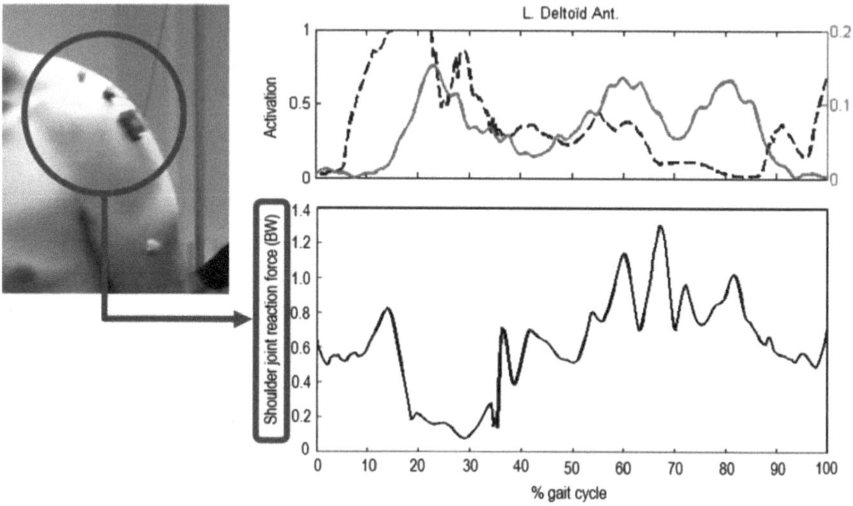

Fig. 2.15 Histories of joint reaction and muscle activation at left shoulder

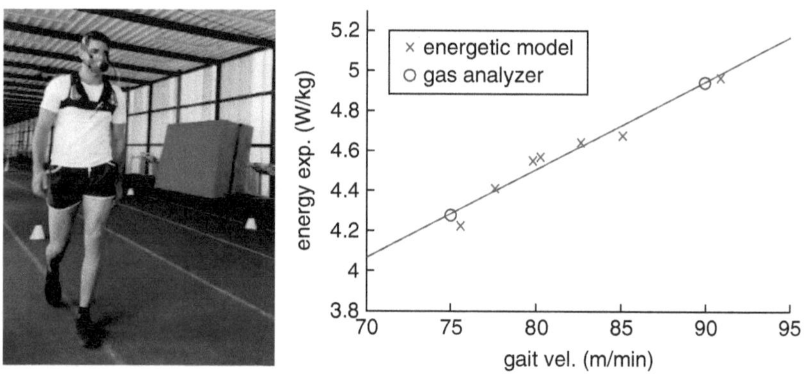

Fig. 2.16 Energy consumption in healthy gait: test and results

the final results were those shown in Fig. 2.16, which offer a good correlation. Therefore, the mentioned model will be applied to SCI subjects to compare energetic cost of the gait when wearing the passive and the active orthoses, thus avoiding the patients to undergo the long and tiring tests carrying the gas analyzer.

The estimation of joint reactions and energetic cost based on motion/force/EMG capture and musculo-skeletal models will hence provide indicators to assess up to what extent the new active orthoses may be more beneficial in the long term for the patient's health than the traditional passive devices.

2.5 Conclusions and Future Work

Low-cost active knee-ankle-foot orthoses can be extremely useful as assistive devices for the gait of SCI subjects possessing some hip flexion, enabling them to walk, generally with the help of crutches, and, hence, to avoid the consequences of sedentarism caused by a systematic use of the wheelchair. Conventional passive knee-locking uprights can be easily and inexpensively converted into smart exoskeletons that detect user's motion intention and provide knee flexion-extension function, enabling a more natural and comfortable gait.

In this work, it has been reported how skeletal models fed with the results of motion/force captures of the patient's gait wearing his/her passive devices can help to customize their active counterparts. Moreover, they were found to be also helpful for the adaptation process of the patient to using the exoskeletons by providing objective parameters that allow the monitoring of the progress achieved during the training sessions, as well as the evaluation of the final advantage provided by the active assistive devices with respect to the passive ones.

Finally, it has been foreseen that musculo-skeletal models receiving the input from motion/force/EMG captures of the patient's gait wearing his/her passive or active devices, can make it possible to compare the improvement attained with the smart orthoses in aspects that may impact the patient's health in the long-term, as joint reactions and energetic cost. Such a comparison will be addressed in the near future.

Compliance with Ethical Standards Research involving human participants: This study was approved by the institutional ethical committee.

Informed Consent All participants in the experiments gave their informed consent.

Funding This work was funded by the Spanish MINECO under project DPI2015-65959-C3-1-R, cofinanced by the EU through the EFRD program, and by the Galician Government under grant ED431B2016/031.

Conflict of Interest The authors have no conflict of interest.

References

1. Bickenbach, J., Bodine, C., Brown, D., Burns, R., Campbell, D., Cardenas, S., Charlifue, S., Chen, Y., Gray, D., Li, L., Officer, A., Post, M., Shakespeare, T., Sinnott, A., von Groote, P., Xiong, X.: International Perspectives on Spinal Cord Injury. World Health Organization, Geneva (2013)
2. Krebs, H.I., Volpe, B.T.: Rehabilitation robotics. Handb. Clin. Neurol. **110**, 283–294 (2013)
3. Yakimovich, T., Lemaire, E.D., Kofman, J.: Engineering design review of stance-control knee-ankle-foot orthoses. J. Rehabil. Res. Dev. **46**(2), 257–267 (2009)
4. Beil, J., Perner, G., Asfour, T.: Design and control of the lower limb exoskeleton KIT-EXO-1. In: Proceedings IEEE International Conference on Rehabilitation Robotics, pp. 119–124. Singapore (2015)

5. Sawicki, G., Ferris, D.: A pneumatically powered knee-ankle-foot orthosis (KAFO) with myoelectric activation and inhibition. J. Neuroeng. Rehabil. **6**(23), 1–16 (2009)
6. Lugris, U., Carlin, J., Luaces, A., Cuadrado, J.: Gait analysis system for spinal cord-injured subjects assisted by active orthoses and crutches. J. Multi-Body Dyn. **227**(4), 363–374 (2013)
7. Garcia de Jalon, J., Bayo, E.: Kinematic and Dynamic Simulation of Multibody Systems. Springer, New York (1994)
8. Kuo, A.D.: A least-squares estimation approach to improving the precision of inverse dynamics computations. J. Biomech. Eng. **120**, 148–159 (1998)
9. Van den Bogert, A.J., Su, A.: A weighted least squares method for inverse dynamic analysis. Comput. Methods Biomech. Biomed. Engin. **11**(1), 3–9 (2008)
10. Rimer, R., Hsiao-Wecksler, E.T.: Improving net joint torque calculations through a two-step optimization method for estimating body segment parameters. J. Biomech. Eng. **131**, 011007-1–011007-7 (2009)
11. Michaud, F., Lugris, U., Ou, Y., Cuadrado, J., Kecskemethy, A.: Influence of muscle recruitment criteria on joint reaction forces during human gait. In: Proceedings ECCOMAS Thematic Conference Multibody Dynamics, paper 141. Barcelona (2015)
12. Michaud, F., Lugris, U., Ou, Y., Cuadrado, J., Kecskemethy, A.: Optimization methods for identifying muscle forces in spinal cord injury subject during crutch gait. In: Proceedings ECCOMAS Thematic Conference Multibody Dynamics, paper 199. Prague (2017)
13. Umberger, B.R., Gerritsen, K.G.M., Martin, P.E.: A model of human muscle energy expenditure. Comput. Methods Biomech. Biomed. Eng. **6**(2), 99–111 (2003)

Chapter 3
Comparison and Analysis of Multibody Dynamics Formalisms for Solving Optimal Control Problem

Quentin Docquier, Olivier Brüls, and Paul Fisette

Abstract Optimal Control methods are increasingly used for the control of multibody systems (MBS). This work analyzes the different dynamic formulations and compare their performances in solving Optimal Control Problem. The focus is on minimal coordinates and the derivation of the dynamics via the recursive methods for tree like MBS (i.e., the so-called Newton-Euler and Order-N recursive algorithms). The different formulations are introduced and their derivations are discussed. A benchmark case study (i.e., a 3D series manipulator balancing an inverted pendulum) is modeled and a series of manipulation tasks (movement of the end effector in the 3D space) are performed. The OCP is formulated and solved with the help of the CasADi software while the dynamic formulations are generated by the Robotran software. Results show that the implicit and semi-explicit formulations derived via the Newton-Euler recursive algorithm lead to faster computation of the OCP than the explicit formulations. This is explained by a more compact expression for the implicit dynamics. However, a lower number of high local minima is observed with the explicit formulations for the most extreme robot manipulations.

Keywords Multibody system dynamics · Direct optimal control · Newton-Euler recursive algorithm · Tree-like system · 3D serial robot · Inverted pendulum

3.1 Introduction

Nowadays, Optimal Control (OC) methods are increasingly used for the control of mechanical systems. Many industrial applications already make use of it, e.g., for the trajectory planning of industrial robots, while the scientific community continues

Q. Docquier (✉) · P. Fisette
Université Catholique de Louvain, Louvain-la-Neuve, Belgium
e-mail: quentin.docquier@uclouvain.be; paul.fisette@uclouvain.be

O. Brüls
Université de Liège, Liège, Belgium
e-mail: o.bruls@uliege.be

E. Zahariev, J. Cuadrado (eds.), *IUTAM Symposium on Intelligent Multibody Systems – Dynamics, Control, Simulation*, IUTAM Bookseries 33,
https://doi.org/10.1007/978-3-030-00527-6_3

to investigate new possibilities in the fields of active orthoses and autonomous driving. Optimal Control of multibody systems (MBS) is of particular interest for many mechanical systems such as vehicles, robots or the human body as it offers the possibility to compute beforehand a control sequence on a given time interval such that a certain optimality criterion is achieved. This makes OC methods suitable for multibody systems whose dynamics involves non-linearities and rotations in the tridimentional space. Moreover, the computation of control command inputs via optimization enables the control of underactuated MBS characterized by a lower number of inputs than degrees of freedom. However, the performances of the optimized control inputs strongly depend on the mathematical model used to represent the dynamics of the controlled system.

Optimal Control is also essential for Model Predictive Control (MPC) of non-linear systems. MPC consists in determining the current control inputs by solving an Optimal Control Problem (OPC) on a finite future time horizon. In case of mechanical systems with fast dynamics, fast updates of the control inputs are required, which means that the OCP online solving must be carried out in real-time. This leads to many investigations on the OCP solving efficiency including for multibody control purpose.

Optimal Control of multibody systems has already been approached on many different angles. There have been a multitude of works to develop general purpose algorithms for optimal trajectory planning of MBS. Among them, studies have focused on *constrained* MBS characterized by loops of bodies and algebraic equations [12]. For such MBS, different methods of index reduction for the differential algebraic equations (DAE) system have been investigated (e.g., coordinate partitioning [21] and Baumgarte stabilization [3]). Such methods are also of great importance in the case of non-minimal parameterization of the system motion. In [11] and [14], different parameterizations of the special orthogonal group SO(3) have been investigated with special integrators adapted to the parametrization. These non-minimal parametrizations are of particular interest for many applications (e.g., flying vehicles such as drones) as they offer a singular-free representation for the rotations.

On the other hand, minimal coordinates have also drawn the attention of researchers as they are appropriate for the representation of many industrial applications (e.g., serial manipulators). During the two past decades, the optimal trajectory planning of industrial robots has been a very popular research subject [2, 6, 18, 19]. These robots generally possess a tree-like structure (i.e., without loop of bodies) which leads to *unconstrained* systems in the case of a joint coordinates approach.

In robotics,the trajectory planning is usually referred as the inverse dynamics problem for which the objective is to determine the joint torques for a prescribed motion of the controlled system [2]. The use of optimization methods is often necessary for the control of underactuated or overactuated systems for which there is still a latitude in the way to actuate them. Among the studied problems, the case of non-minimal phase systems represents an additional challenge as the inverse

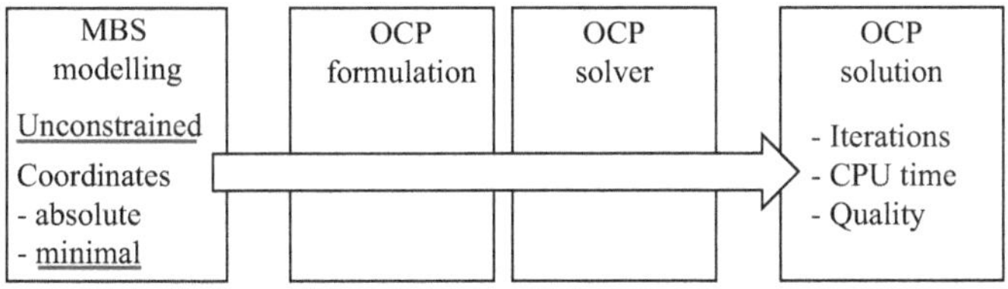

Fig. 3.1 In this work, the focus is on the influence of the MBS modeling on the solving OCP performances

dynamics is, in that case, non-causal which means that the control should start before the beginning of the trajectory [2].

Generally, the research on OC for MBS has mainly focused on the development of new methods for Numerical Optimal Control that are based on a given formulation of the multibody dynamics [2, 5, 19] and [17]. Nevertheless, in [6], Diehl gives a good overview on the different dynamic formulations and how they could influence the Optimal Control solver. He suggests that an implicit formulation of the dynamics through the Newton-Euler Recursive scheme should lead to a lower evaluation cost per iteration. However, to our knowledge, there has been no thorough investigation on the dynamic formulations and its influence on the solving of optimal control problem for *unconstrained* multibody systems. Regarding the numerical formulation of OCP, two methods exist: the multiple shooting and the direct collocation [4]. For the multiple shooting, the integration and the optimization are treated separately. The choice of the integrator is left free and can be adapted to the system dynamics. On the other hand, the direct collocation method consists in imposing the dynamics equations at intermediate collocation points. The integration and the optimization are treated at the same level. The advantages of collocation methods are that they lead to a very sparse NLP and that they show fast local convergence. In addition, they treat unstable system well and can easily cope with state and terminal constraints [6]. These reasons led us to consider the collocation method in the case study of this paper (see Sect. 3.4).

In this work, we analyze different multibody dynamic formulations and compare their performances in terms of convergence, accuracy and computation cost in the OCP framework. The aim is not to develop new numerical methods for the formulation of the OCP or to propose a new type of solver but to provide some insights on existing MBS formulations and their suitability in the formulation of Optimal Control Problems (see Fig. 3.1). Among the different MBS formalisms, we focus on minimal (i.e., relative) coordinates and the derivation of the dynamics via the recursive methods for tree-like MBS (i.e., the so-called Newton-Euler and Order N recursive algorithms). The derivation is done symbolically through the use of the Robotran software [7]. The symbolic equation generator of Robotran has already been proved to be particularly efficient in the derivation of compact expressions for the dynamics of multibody system [15]. However, its performances for OCP solving have never been thoroughly analyzed.

In order to compare the formulations, a benchmark has been developed. It consists of a 3D serial robot arm with an inverted pendulum attached to the end effector. An important point is the underactuated nature of the system (due to the free rotation of the pendulum) which requires the use of optimization to determine the robot control inputs. In addition, the balancing of an inverted pendulum is a well-known task that has already been largely investigated. The system being tridimensional, its non-linear and complex dynamics should highlight the differences between the considered dynamic formulations.

This chapter is organized as follows: first the theory related to the formulation of the OCP and the multibody modeling are respectively presented in Sects. 3.2 and 3.3. The different formulations are introduced and their derivations are discussed. Then, Sect. 3.4 presents a benchmark case study for an *unconstrained* multibody system. First the MBS system (i.e., a 3D robot arm with an inverted pendulum) is described, then the optimal control task is defined. To this extent, the CasADi software [1] is used to formulate and solve the OCP while the dynamic formulations are provided by the symbolic MBS software Robotran [7]. Finally, results are presented for a series of tasks realized with the different representations of the dynamics. These are compared in terms of solver performances: number of iterations, CPU time to convergence and quality of the solutions. Finally, the differences observed for the different formulations are discussed in light of theoretical considerations.

3.2 Optimal Control of Multibody Systems

There are many ways to formulate and solve an Optimal Control Problem. In this Section, we introduce the main principle and define the trajectory planning problem to solve. In Sect. 3.2.1, the continuous problem is presented analytically then the discrete formulation is introduced in Sect. 3.2.2.

3.2.1 Optimal Control Formulation

In a multibody context, the optimal Control Problem (OCP) is formulated as an optimization problem whose solution provides the control trajectory $\mathbf{u}(t)$ that minimizes a given cost function (3.1) and steers the state $\mathbf{x}(t)$ of the multibody system and its algebraic variables $\mathbf{z}(t)$ from a given initial state (3.2) to a given terminal state (3.3), within a given time horizon T (see Fig. 3.2). The OCP then reads:

$$\underset{\mathbf{x}(t),\mathbf{u}(t),\mathbf{z}(t)}{\text{minimize}} \int_0^T L\left(\mathbf{x}(t), \mathbf{u}(t), \mathbf{z}(t)\right) dt \tag{3.1}$$

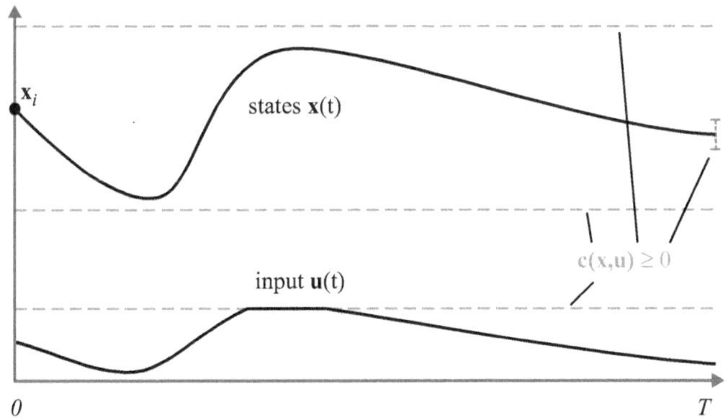

Fig. 3.2 The continuous optimal control problem consists in finding the optimal state and control trajectory on the time interval $[0, T]$

subject to possible path constraints:

$$\mathbf{x}(0) - \mathbf{x}_i = \mathbf{0} \tag{3.2}$$

$$\mathbf{x}(T) - \mathbf{x}_d = \mathbf{0} \tag{3.3}$$

$$\mathbf{c}\left(\mathbf{x}(t), \mathbf{u}(t), \mathbf{z}(t)\right) \geq \mathbf{0} \quad t \in [0\ T] \tag{3.4}$$

The trajectory $\mathbf{x}(t), \mathbf{z}(t)$ and the control inputs $\mathbf{u}(t)$ are subjected to equality constraints denoted $\mathbf{g}$, which typically represent the system dynamics expressed as a set of Differential Algebraic Equations (DAE).

$$\mathbf{g}\left(\mathbf{x}(t), \mathbf{u}(t), \mathbf{z}(t)\right) : \begin{cases} \dot{\mathbf{x}} = \mathbf{f}(\mathbf{x}, \mathbf{u}, \mathbf{z}) \\ \mathbf{a}(\mathbf{x}, \mathbf{z}) = \mathbf{0} \end{cases} \quad t \in [0\ T] \tag{3.5}$$

In this equation, $\mathbf{f}$ and $\mathbf{a}$ respectively denote the differential and the algebraic equations that described the physics of the system in terms of the states $\mathbf{x}$, the algebraic variables $\mathbf{z}$ and the control inputs $\mathbf{u}$. This general framework for the dynamics is at the core of the CasADi OC program and its permits the modeling of the dynamics of the multibody system as a system of DAE (see Sect. 3.3). The optimization problem defined by the set of Eqs. (3.1), (3.2), (3.3), (3.4) and (3.5) has in general no analytical solution. However it can be solved numerically through its discretization on the time interval $[0, T]$.

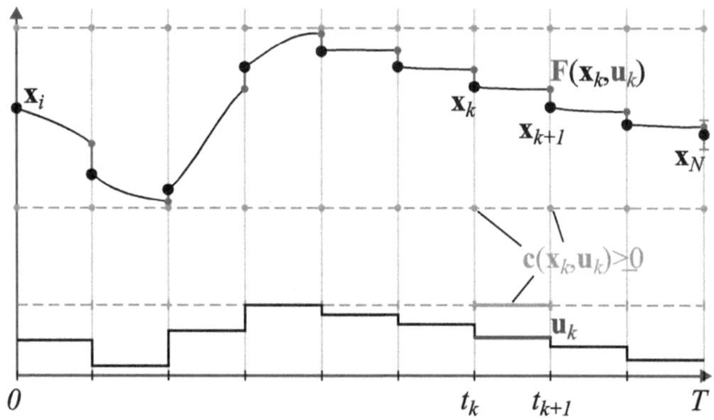

Fig. 3.3 Numerical OCP with the dynamic constraints (3.10) that are unsatisfied. The decision variables are the states at the discrete points and the piecewise constant control inputs

3.2.2 Numerical Optimal Control Methods

There are several approaches to obtain a numerical solution of an OCP. Among them, the *direct methods* consist in discretizing the time horizon $[0, T]$ into N time intervals $[t_k, t_{k+1}]$ such that $t_0 = 0$, $t_N = T$ and $k \in 0, 1, \ldots, N-1$. The time functions $\mathbf{x}(t)$, $\mathbf{u}(t)$ and $\mathbf{z}(t)$ that were the unknowns of the system (3.1), (3.2), (3.3), (3.4) and (3.5) are replaced by a sequence of $N+1$ discrete states $\mathbf{x}_k$, $\mathbf{z}_k$ and N piecewise constant controls $\mathbf{u}_k$ which are now the decision variables of the optimization problem in its discrete form (see Fig. 3.3).

The cost function (3.1) is replaced by a numerical approximation p which is a function of the decision variables vector $\mathbf{w} = \{\mathbf{x}_0, \mathbf{u}_0, \mathbf{z}_0, \ldots, \mathbf{x}_k, \mathbf{u}_k, \mathbf{z}_k \ldots \mathbf{u}_{N-1}, \mathbf{z}_{N-1}, \mathbf{x}_N\}$. Finally, the path constraints (3.2) and (3.3) are re-expressed in terms of the decision variables vector $\mathbf{w}$. This parameterization of the original continuous OCP is referred as a nonlinear program (NLP).

$$\underset{\mathbf{w}}{\text{minimize}}\, p(\mathbf{w}) \tag{3.6}$$

subject to path constraints on the decision variables:

$$\mathbf{x}_0 - \mathbf{x}_i = \mathbf{0} \tag{3.7}$$

$$\mathbf{x}_N - \mathbf{x}_d = \mathbf{0} \tag{3.8}$$

$$\mathbf{c}\,(\mathbf{w}) \geq \mathbf{0} \tag{3.9}$$

In order to respect the system dynamics, the constant control input $\mathbf{u}_k$ on the time interval $[t_k, t_{k+1}]$ is linked to the two adjacent differential states $\mathbf{x}_k$ and $\mathbf{x}_{k+1}$ through the explicit numerical integration $\mathbf{F}$ of the continuous dynamics (3.5) on the same discrete time interval (see Fig. 3.3).

$$\mathbf{x}_{k+1} - \mathbf{F}(\mathbf{x}_k, \mathbf{u}_k, \mathbf{z}_k) = 0 \tag{3.10}$$

In addition, the differential and algebraic states are linked at each discrete time through the set of algebraic equations.

$$\mathbf{a}(\mathbf{x}_k, \mathbf{z}_k) = 0 \tag{3.11}$$

As mentioned in the introduction, a collocation method is used to deal with the constraints (3.10) and (3.11). This technique consists in defining collocation points between t_k and t_{k+1} at which the states, represented as polynomial functions, have to satisfy the continuous dynamics (3.5). The integration scheme implies the definition of extra variables (i.e., the polynomial coefficients) and additional constraint equations. This leads to a very large but sparse NLP.

The optimization problem defined by Eqs. (3.6), (3.7), (3.8), (3.9), (3.10) and (3.11) can be solved with different techniques (e.g., an Interior Point method) which generally require the derivatives of the cost function and constraint equations with respect to the decision variables. To this extent, the software CasADi [1] provides sensitivity analysis tools for ODE and DAE system that allows the automatic generation of the derivatives necessary to solve the NLP. To do so, CasADi uses state-of-the-art algorithmic differentiation [10].

3.3 Multibody Systems Modeling

It is well-established that several formalisms can be used to model multibody systems (MBS). Among them, both the absolute and relative coordinates can be considered for optimal control, resulting in different structures for the NLP to solve [2]. In this study, the focus is on minimal coordinates and the different formulations of the dynamics. First the relative (i.e., minimal) coordinates approach is briefly described. Then the different forms for the MBS dynamics that can be obtained through recursive algorithms are presented. Finally, the symbolic generation of these dynamics equations is discussed.

3.3.1 *Relative Coordinates Approach*

In the relative coordinates approach, the parametric configuration of a body is defined relatively to another one called its parent body. The motion degrees of freedom (dof) are defined via the joints that connect the different bodies of the system. If the MBS does not contain any loops of bodies, it inherits a tree-like structure as each body can only have one parent body (see Fig. 3.4). The number of coordinates is minimal and thus reduced compared to the absolute coordinates approach, as the joint constraints are implicitly taken into account by the relative formulation, leading to a pure set of ordinary differential equations.

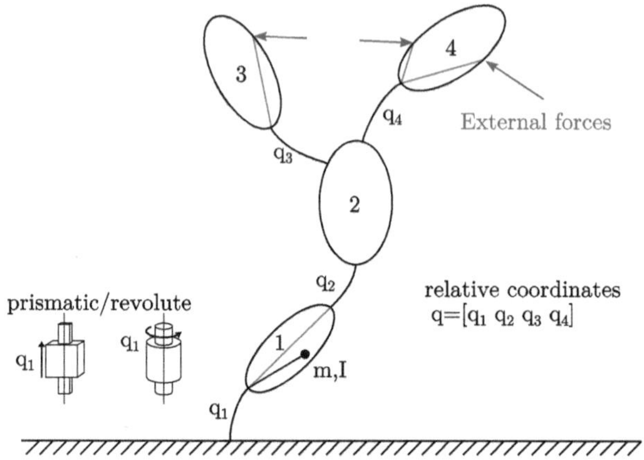

Fig. 3.4 The relative coordinates approach consists in defining the motion of each body relatively to its parent body with the help of a joint coordinate q

In the relative coordinates approach, a MBS is fully characterized by two types of features:

- bodies, defined by their mass, center of mass and inertia tensor (10 parameters in total),
- joints, defined by their location on the body and their nature (i.e., prismatic or revolute, free or constrained).

A tree-like structure might be quite restrictive since many mechanical systems contain kinematic loops (e.g., a four-bar linkage). The relative coordinates approach deals with such closed-loop systems by introducing loop constraint equations. The tree-like structure is restored in two steps. First the loop is cut resulting into two independent branches of bodies. Afterwards the loop closure is ensured by applying algebraic constraints, for instance forcing two points of the created branches to coincide or imposing a given distance between two points. Hence, the derivation of the *differential* equations is based on the tree-like structure of the MBS whether or not the system is constrained. In this work, only unconstrained tree-like MBS are considered within the frame of OCP formulation. However, as the derivation of the differential equations is the same for constrained system, the study of the different formulations of the dynamics will be insightful for both constrained and unconstrained MBS.

3.3.2 *Formulations of the Dynamic Equations*

In this section, different formulations of the equations of motion are presented for the case of *unconstrained* MBS. Among those formulations, the so-called *Newton-Euler recursive* scheme denoted NER is particularly suitable for minimal coordinates [13]. This method implements two consecutive recursions on the MBS (a forward kinematics followed by a backward dynamics). It is well-known for its facility of computer implementation and its reduced number of operations necessary to obtain the equations of motion. It expresses the vector of generalized forces $\mathbf{Q}$ as a function of the system kinematics $(\mathbf{q}, \dot{\mathbf{q}}, \ddot{\mathbf{q}})$.

$$\boldsymbol{\Phi}(\ddot{\mathbf{q}}, \dot{\mathbf{q}}, \mathbf{q}) = \mathbf{Q} \tag{3.12}$$

This very efficient formulation, also called inverse dynamics has only a $\mathcal{O}(N)$ complexity thanks to its recursive formulation, N being the size of the MBS (i.e., the number of joints). It is extensively used by roboticians to determine the actuation torques $\mathbf{Q}$ necessary to track a given trajectory $(\mathbf{q}(t), \dot{\mathbf{q}}(t), \ddot{\mathbf{q}}(t))$ for a fully actuated robot.

3.3.2.1 Implicit Formulation

More generally, Eq. (3.12) can be used to represent the dynamics of any MBS systems, in a residual form. In that case, the joint torques/forces $\mathbf{Q}$ can be configuration-dependent (i.e., an elastic joint or a feedback control law) and are generally expressed as a function of the generalized coordinates $\mathbf{q}$ and generalized velocities $\dot{\mathbf{q}}$. It results in an implicit formulation of the equations of motion in terms of the generalized acceleration $\ddot{\mathbf{q}}$:

$$\boldsymbol{\Phi}(\ddot{\mathbf{q}}, \dot{\mathbf{q}}, \mathbf{q}) - \mathbf{Q}(\dot{\mathbf{q}}, \mathbf{q}) = \mathbf{0} \tag{3.13}$$

According to Sect. 3.2.1, some control inputs $\mathbf{u}(t)$ are considered for the OCP formulation. To this extent, the joint forces $\mathbf{Q}$ can be partitioned into two subsets.

- $\mathbf{Q_a}$ for the active or actuated joints (input of the system)
- $\mathbf{Q_p}$ for the passive joints that can contain dry friction, damping force, etc.

$$\mathbf{Q}(\dot{\mathbf{q}}, \mathbf{q}) = \mathbf{Q_a} + \mathbf{Q_p}(\dot{\mathbf{q}}, \mathbf{q}) \tag{3.14}$$

The system (3.13) can be re-expressed as a DAE system of the form (3.5) with the generalized coordinates and velocities representing the differential states and the accelerations considered as the algebraic states.

$$\mathbf{x} = \begin{bmatrix} \dot{\mathbf{q}} \\ \mathbf{q} \end{bmatrix} \qquad \mathbf{z} = \ddot{\mathbf{q}} \qquad \mathbf{u} = \mathbf{Q_a}$$

$$\begin{cases} \mathbf{f}(\mathbf{x}, \mathbf{u}, \mathbf{z}) = \begin{bmatrix} \mathbf{z} \\ \dot{\mathbf{q}} \end{bmatrix} \\ \mathbf{a}(\mathbf{x}, \mathbf{z}) \quad = \boldsymbol{\Phi}(\mathbf{q}, \dot{\mathbf{q}}, \mathbf{z}) - \mathbf{Q}(\mathbf{q}, \dot{\mathbf{q}}, \mathbf{u}) = \mathbf{0} \end{cases} \tag{3.15}$$

3.3.2.2 Semi-explicit Formulation

The original Newton-Euler recursive scheme leading to the formulation (3.12) can be modified in order to obtain a semi-explicit form of the dynamics equations for tree-like multibody systems. This new form still preserves the full recursivity even for the computation of the mass matrix [9].

$$\mathbf{M}(\mathbf{q})\ddot{\mathbf{q}} + \mathbf{c}(\mathbf{q}, \dot{\mathbf{q}}) = \mathbf{Q}(\dot{\mathbf{q}}, \mathbf{q}) \tag{3.16}$$

$\mathbf{M}$ is the symmetric generalized mass matrix of the system and $\mathbf{c}$ is the non-linear dynamic vector which contains the gyroscopic, centripetal and gravity terms as well as the contribution of the resultant external forces. The difference compared to Eq. (3.12) is that the algorithm first computes $\mathbf{M}$ and $\mathbf{c}$ individually before the evaluation of the left-hand side.

This semi-explicit formulation is often preferred for time integration purposes. However, the factorization of the mass matrix in the recursive scheme requires additional computational efforts of $\mathcal{O}(N^2)$ complexity [9]. The semi-explicit system (3.16) can be re-expressed as a DAE system in a similar way as for the implicit formulation.

$$\mathbf{x} = \begin{bmatrix} \dot{\mathbf{q}} \\ \mathbf{q} \end{bmatrix} \qquad \mathbf{z} = \ddot{\mathbf{q}} \qquad \mathbf{u} = \mathbf{Q_a}$$

$$\begin{cases} \mathbf{f}(\mathbf{x}, \mathbf{u}, \mathbf{z}) = \begin{bmatrix} \mathbf{z} \\ \dot{\mathbf{q}} \end{bmatrix} \\ \mathbf{a}(\mathbf{x}, \mathbf{z}) \quad = \mathbf{M}(\mathbf{q})\, \mathbf{z} + \mathbf{c}(\mathbf{q}, \dot{\mathbf{q}}) - \mathbf{Q}(\dot{\mathbf{q}}, \mathbf{q}, \mathbf{u}) = \mathbf{0} \end{cases} \tag{3.17}$$

3.3.2.3 Explicit Formulation

The system of Eq. (3.16) can be solved in terms of the accelerations $\ddot{\mathbf{q}}$ via a Cholesky decomposition of the mass matrix leading to the explicit expression $\boldsymbol{\gamma}$ for the acceleration $\ddot{\mathbf{q}}$ in terms of the position $\mathbf{q}$ and velocity $\dot{\mathbf{q}}$. This resolution step yields an additional $\mathcal{O}(N^3)$ complexity compared to the semi-explicit form.

$$\ddot{\mathbf{q}} = \mathbf{M}^{-1}\, [\mathbf{Q} - \mathbf{c}]\, (\dot{\mathbf{q}}, \mathbf{q}) \triangleq \boldsymbol{\gamma}(\dot{\mathbf{q}}, \mathbf{q}, \mathbf{Q}) \tag{3.18}$$

Another way to derive the explicit form of the equations of motion in a fully recursive way can be obtained through the Order-N algorithm [16]. This method is based on three recursive steps (instead of two for the NER scheme) on the multibody system. In short, first a forward recursion covers the MBS from the inertial body to the leaves to compute the bodies *kinematics* (position, orientation, linear and angular velocities). A second backward recursion computes the bodies *dynamics* on the basis of the classical Newton-Euler equations within which a subtle embedded block factorization of the mass matrix is performed. This process prepares the third final *kinetic* recursion that straightforwardly computes the generalized acceleration $\ddot{\mathbf{q}}$ as a function $\boldsymbol{\nu}$ of the joint positions and velocities. This three-steps algorithm has a $\mathcal{O}(N)$ complexity. Several implementations of it have been proposed in the literature, the one used in this paper is the Schwertassek-Rulka method [16].

$$\ddot{\mathbf{q}} = \boldsymbol{\nu}(\dot{\mathbf{q}}, \mathbf{q}, \mathbf{Q}) \tag{3.19}$$

As the system is expressed in a purely explicit form, the accelerations $\ddot{\mathbf{q}}$ do not *necessarily* need to be considered as an algebraic state. As a result, the dynamics can be assimilated to either set of ODE or DAE. The explicit dynamics is generally denoted $\boldsymbol{\xi}$ and can either represent the NER explicit dynamics $\boldsymbol{\gamma}$ or the Order-N dynamics $\boldsymbol{\nu}$. The dynamics is represented by a set of differential algebraic equations (DAE) as

$$\mathbf{x} = \begin{bmatrix} \dot{\mathbf{q}} \\ \mathbf{q} \end{bmatrix} \qquad \mathbf{z} = \ddot{\mathbf{q}} \qquad \mathbf{u} = \mathbf{Q_a}$$

$$\begin{cases} \mathbf{f}(\mathbf{x}, \mathbf{u}, \mathbf{z}) = \begin{bmatrix} \mathbf{z} \\ \dot{\mathbf{q}} \end{bmatrix} \\ \mathbf{a}(\mathbf{x}, \mathbf{z}) \quad = \mathbf{z} - \boldsymbol{\xi}(\dot{\mathbf{q}}, \mathbf{q}, \mathbf{u}) = \mathbf{0} \end{cases} \tag{3.20}$$

or by a set of ordinary differential equations (ODE) as

$$\mathbf{x} = \begin{bmatrix} \dot{\mathbf{q}} \\ \mathbf{q} \end{bmatrix} \qquad \mathbf{u} = \mathbf{Q_a}$$

$$\mathbf{f}(\mathbf{x}, \mathbf{u}) = \begin{bmatrix} \boldsymbol{\xi}(\dot{\mathbf{q}}, \mathbf{q}, \mathbf{u}) \\ \dot{\mathbf{q}} \end{bmatrix} \tag{3.21}$$

3.3.3 Symbolic Generation

In practice, the MBS equations of motion can be computed either numerically or symbolically. In specific cases, the recursive nature of the relative coordinates can be deeply exploited by a symbolic program dedicated to multibody systems. This is the purpose of the Robotran software [7] that offers symbolic tools for an extensive simplification of the equations of motion. The latter are written under the form of a function (C, Matlab, Python) with adequate input/output arguments. These functions, internally optimized thanks to the symbolic approach, can be used externally as a black-box by another program or computer environment.

Table 3.1 Formulations for the dynamics equations of *unconstrained* multibody systems

	Algorithm	Section	Dynamics	Complexity
IMPL	Original NER	3.3.2.1	$\mathbf{\Phi} - \mathbf{Q} = \mathbf{0}$	N
SEMI	Modified NER	3.3.2.2	$\mathbf{M\ddot{q}} + \mathbf{c} - \mathbf{Q} = \mathbf{0}$	N^2
EXP1	Modified NER + Cholesky	3.3.2.3	$\mathbf{\ddot{q}} - \boldsymbol{\gamma} = \mathbf{0}$	$N^2 + N^3$
EXP2	Order-N	3.3.2.3	$\mathbf{\ddot{q}} - \boldsymbol{\nu} = \mathbf{0}$	N

In this paper, the function generated by Robotran will be used by the CasADi software [1] to express the system dynamics for the OCP formulation.

The different formulations of the dynamics yield different levels of complexity as summarized in Table 3.1.

3.4 Case Study: Unconstrained MBS

Given their different complexities, the formulations of the dynamics should lead to distinct performances for the OCP solving. To this extent, the Optimal Control of a simple unconstrained multibody system has been extensively tested.

3.4.1 System Presentation

The system of interest is a 3D robot arm with an inverted pendulum attached to the end effector as represented in Fig. 3.5. The 3D arm has a morphology that resembles the human arm. It is composed of 3 bodies and 4 actuated rotational joints (with relative coordinates q_1, q_2, q_3 and q_4). The first element is connected to the base with two rotational joints enabling to point in any direction of space. Then the second and third elements are connected in series via a single rotational joint. The relative motion of the pendulum with respects to the end effector is characterized by two Cardan-type angles (q_5 and q_6) which are considered as free joints (i.e., frictionless and non-actuated). The MBS is thus composed of 4 bodies and 6 joints. The system is underactuated as only the 4 joints of the arm are actuated. The mass, inertia and length of the different bodies are given in Table 3.2.

3.4.2 OCP Formulation

There are many possible ways to formulate the Optimal Control Problem for this application. In this study, the focus is on the trajectory optimization of the 3D arm. The task is to move the system from an initial pose and position its end effector at a given location in the 3D space after an imposed time T. At this final time, the system must be at equilibrium with the pendulum in a perfect upright position.

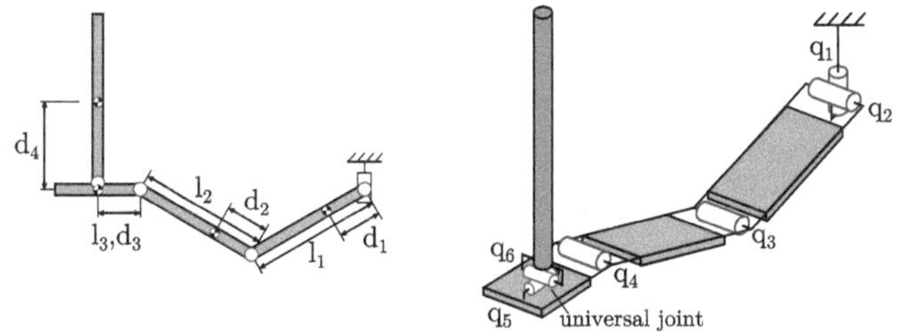

Fig. 3.5 Geometrical representation of the studied MBS system

Table 3.2 Geometric and inertia parameters for the 3D serial robot and the pendulum

	Length [m]	c.o.m. position [m]	Mass [kg]	Inertia [kg m^2]
Upperarm	$l_1 = 0.3$	$d_1 = 0.1$	2.5	[0.06,0.1,0.1]
Forearm	$l_2 = 0.3$	$d_2 = 0.1$	2	[0.03,0.05,0.05]
Hand	$l_3 = 0.1$	$d_3 = 0.1$	0.5	[0.004,0.006,0.006]
Pendulum		$d_4 = 0.2$	1	[0.1,0.1,0.05]

c.o.m. stands for the center of mass

The objective function is designed to minimize the actuation via the use of the square of the actuation torques and its time derivative:

$$L(\mathbf{x}, \mathbf{u}, \mathbf{z}) = k_1 \sum_{i=0}^{4} Q_i^2 + k_2 \sum_{i=0}^{4} \dot{Q}_i^{\,2} \tag{3.22}$$

The first term aims at minimizing the torques in the four actuated joints while the second term minimizes the derivative of the controls which smooths the optimal trajectory and limits the mechanical power transmitted into the joints. In practice, the control torques are piece-wise constant on each discrete time interval, which implies the computation of the derivative via finite differences.

To achieve the equilibrium at the final time T, constraints are applied on the differential states $\mathbf{x} = [\dot{\mathbf{q}}\ \mathbf{q}]$, the inputs $\mathbf{u} = \mathbf{Q_a}$ and the algebraic states $\mathbf{z}$ when appropriate.

- The absolute position of the end effector can be expressed as a function of the angular joint positions, $\mathbf{s}_{ee}(\mathbf{q})$. Thus, the end effector can be constrained to a desired absolute Cartesian position $\mathbf{p}_d = [x_d\ y_d\ z_d]$:

$$\mathbf{p}_d - \mathbf{s}_{ee}(\mathbf{q}(T)) = \mathbf{0} \tag{3.23}$$

- The absolute velocity of the end effector being a linear function of the joint angular velocities, equating the angular velocities at the final time T to zero should then be sufficient.

$$\dot{\mathbf{q}}(T) = \mathbf{0} \tag{3.24}$$

- The equilibrium of the whole system is achieved by canceling the joint accelerations at the final time, $\ddot{\mathbf{q}}(T) = \mathbf{0}$. To this extent, the joint accelerations are either available as algebraic variables in the case of a DAE system or expressed with the function $\boldsymbol{\xi}$ in the case of a pure ODE system (see Sect. 3.3.2.3 on explicit formulations).

$$\mathbf{z}(T) = \mathbf{0} \quad \text{or} \quad \boldsymbol{\xi}(\dot{\mathbf{q}}, \mathbf{q}, \mathbf{Q_a}) = \mathbf{0} \tag{3.25}$$

At time $t = 0$, the robot is in a stable equilibrium. The initial condition of the system is given by the joint angles, $\mathbf{q}_i$.

$$\mathbf{q}(0) = \mathbf{q}_i = [0\ ;\ \alpha\ ;\ -2\alpha\ ;\ \alpha\ ;\ 0\ ;\ 0] \quad \dot{\mathbf{q}}(0) = \mathbf{0} \quad \ddot{\mathbf{q}}(0) = \mathbf{0} \tag{3.26}$$

In addition, path inequality constraints (3.27) are applied to the joint angles to keep a physical sense for the arm and the pendulum motions.

$$\begin{aligned} -\infty &\le q_1 \le \infty \\ -\pi/2 &\le q_2 \le \pi/2 \\ -5\pi/12 &\le q_3 \le 0 \\ -\pi/2 &\le q_4 \le \pi/2 \\ -\pi/4 &\le q_5 \le \pi/4 \\ -\pi/4 &\le q_6 \le \pi/4 \end{aligned} \tag{3.27}$$

The equilibrium condition (3.25), the terminal constraint (3.23) and the path constraints ensure all together that the pendulum is in an upright position at the final time.

3.4.3 *Numerical Results*

In order to assess the performances of the different formulations, we have compared the solutions obtained for the above 3D arm benchmark. For the comparison, a total of six formulations for the system dynamics have been considered. From Sect. 3.3, four different formulations for the equations of motion have been derived:

- IMPL: Implicit formulation from the Newton-Euler Recursive scheme (NER),
- SEMI: Semi-explicit formulation from the NER scheme,
- EXP1: Explicit formulation from the NER scheme,
- EXP2: Explicit formulation from the Order-N formalism.

While the first two formulations imply a DAE system for its numerical representation, the two explicit formulations can either be numerically treated as a system of DAE or as a system of pure ODE. The difference lies in the number of decision variables and constraint equations for the numerical NLP but also in the treatment of the terminal constraints.

- EXPi-O: system of ODE with the position and velocity in the differential states,
- EXPi-A: system of DAE with the joint accelerations as algebraic variables.

The OCP described in Sect. 3.4.2 is formulated with the help of the software environment CasADi [1]. Based on the functions generated by Robotran, symbolic expressions of the dynamics are written and symbolically derived with respect to the states and input variables. Then, the numerical OCP is formulated using a direct collocation scheme [4] allowing the treatment of either a DAE or a pure ODE representation of the dynamics. The resulting NLP is then solved using the Interior Point solver IPOPT [20].

3.4.3.1 Simulation Settings

The performances of each formulation are evaluated through the trajectory optimization of the 3D arm robot. The task is described in Sect. 3.4.2. The time interval length T is 2 s and is divided into 100 time intervals for the numerical solving via a direct collocation method.

To compare the formulations, the optimization was performed for a series of end-effector terminal locations. This set of points is defined in cylindrical coordinates (φ,r,z) around a vertical axis passing through the robot arm origin. The angular position φ varies in the range $[-\pi/2, \pi/2]$ rad. On the other hand, the two other coordinates have relatively small variation ranges because the end effector has to remain in the robot working space: the radius is fixed to 0.63 m and the height varies in the range $[-0.1, 0.1]$ m (see Fig. 3.6).

Among the defined lateral surface, many points can be tested (see Fig. 3.6). The number of trajectory optimizations was chosen on the basis of a convergence analysis. The optimizations have been performed on a grid of different sizes and it has been shown that the mean number of iterations does not vary significantly between two refinements after a certain size (see Fig. 3.7). The chosen grid is composed of 21 angular orientations and 11 vertical positions for a total of 231 points in the 3D space (see Fig. 3.6).

In addition, two different sets of results are generated by tuning the k_1 and k_2 coefficients. The first cost function is defined by ($k_1 = 1\text{e}0$, $k_2 = 0$) and corresponds to a more aggressive actuation while the second cost function is defined by ($k_1 = 1\text{e}0$, $k_2 = 1\text{e}5$) and should result in a smoother actuation. In both cases, the initial condition (i.e., the α angle) is such that the end effector is positioned at 0.65 m from the origin with zero lateral rotation of the arm (i.e., $\varphi = 0$ rad).

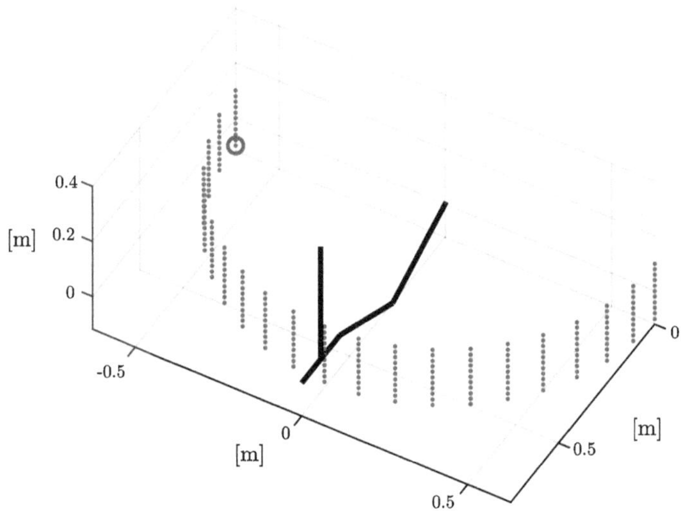

Fig. 3.6 Grid of the 3D point to reach. The multibody system is represented in its initial condition

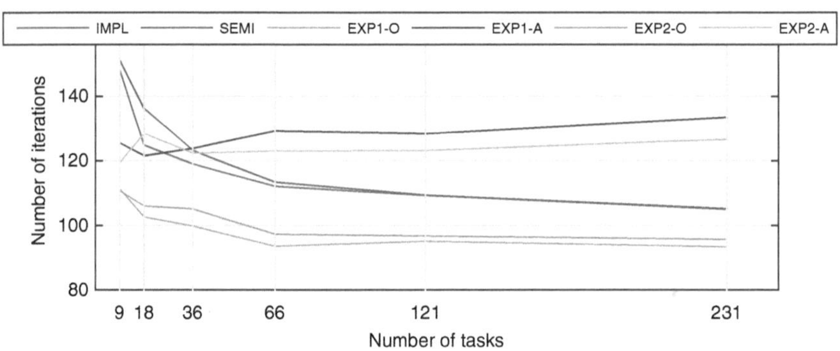

Fig. 3.7 Convergence analysis for the different formulations. After a given number of tasks, the number of iterations does not vary significantly

3.4.3.2 Simulation Results

Figure 3.8 shows the optimal trajectory obtained to reach the point $\mathbf{p}_d = \{0, 0.63, -0.1\}$ (see red point in Fig. 3.6) for the two different definitions of the cost function. The second definition leads to a smoother but more important actuation. The torques variations are reduced and there are less oscillations at the end of the trajectory. This smoother actuation is reflected in the joint velocities with lower values than for the non-smoothed actuation. Finally, for the two cost definitions, the same final angular positions and the same final actuation are obtained as the equilibrium to reach after 2 s is identical.

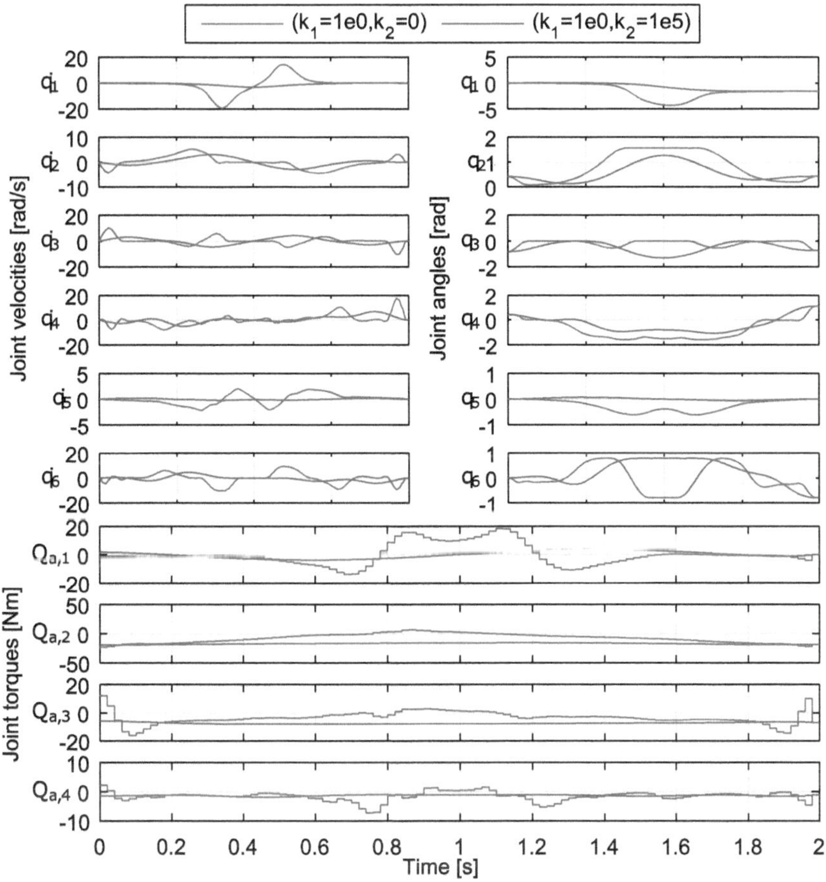

Fig. 3.8 Solution for the two different sets of cost function coefficients. A non-zero coefficient k_2 on the control derivative leads to a smoother but more important actuation

More than the solution itself, the interest of the study is to compare the different formulation performances to solve the OCP described in Sect. 3.4.2. To do so, the number of iterations before convergence, the cost function value at the optimum and the mean evaluation times of the dynamics are evaluated for the two sets of results.

First, the optimal values of the cost function are compared for the six considered representations of the dynamics. Figure 3.9 shows a box-plot[1] of the cost value for the two sets of results (i.e., different coefficients in the cost function). The OCP formulation with the smoothed actuation leads to more constant results with only

[1]On the box plot, the central mark indicates the median, and the bottom and top edges of the box indicate the 25th and 75th percentiles, respectively. The whiskers extend to the most extreme data points not considered outliers, and the outliers are plotted individually using the '+' symbol.

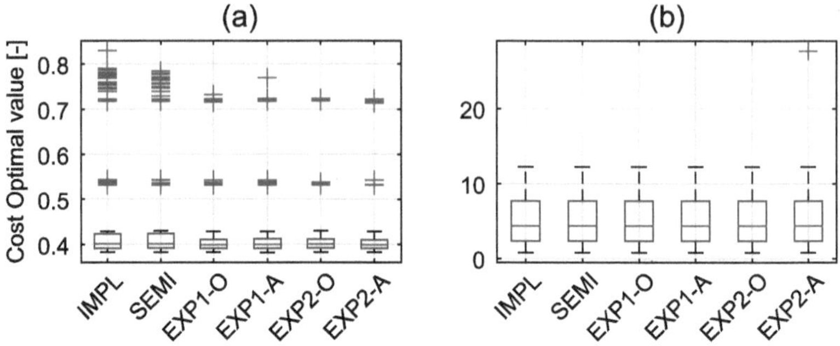

Fig. 3.9 Optimal values of the cost function represented in the form of a box plot for the two sets of solutions. (**a**) First cost function without smoothing: $k_1 = 1e0$, $k_2 = 0$. (**b**) Second cost function with smoothing: $k_1 = 1e0$, $k_2 = 1e5$

one formulation presenting an outlier which corresponds to a higher local minima (see Fig. 3.9b). On the other hand, in the case of the non-smoothed control, the solution leads to a higher number of outliers especially in the case of the implicit and the semi-explicit formulations (see Fig. 3.9a). Note that the higher group of outliers corresponds to the tasks where the point to reach is either at the far left or the far right and the lower group corresponds to the central point (i.e., no lateral movement needed for the arm). The higher outliers clearly correspond to higher local minima as lower minima are reached by the explicit formulations for the same tasks. Regarding the problem with no lateral movement, it is not clear whether it is a local minimum as all the formulations seem to struggle to solve this problem. In general, the actuation smoothing seems to reduce the problem of higher local minima.

Another important aspect of the performances is the computation efforts needed to reach the optimal solution. Figure 3.10 shows the number of iterations to convergence and the CPU time for the two sets of results. Figure 3.11 shows the CPU time per iteration. For the first set (see Fig. 3.10a), the number of iterations is characterized by a high number of outliers corresponding to the high local minima (see Fig. 3.9a). On this same set, the DAE explicit formulations result in a larger average number of iterations than the four other formulations. For the second set of tasks, the implicit and semi-explicit formulation generally perform better than the explicit ones (see Fig. 3.10b). This difference could be due to the high local minima that are encountered by the first two formulations while solving the first set of tasks. Regarding the CPU time, the implicit and the semi-explicit formulations lead to better results than the explicit ones for both sets of tasks. This tendency is also observed for the CPU time per iteration (see Fig. 3.11).

Regarding the numerical formulations, the ODE representation gives better results in terms of iterations for the first set of tasks (see Fig. 3.10a) but this difference is not so clear for the second set of tasks (see Fig. 3.10b). The two explicit

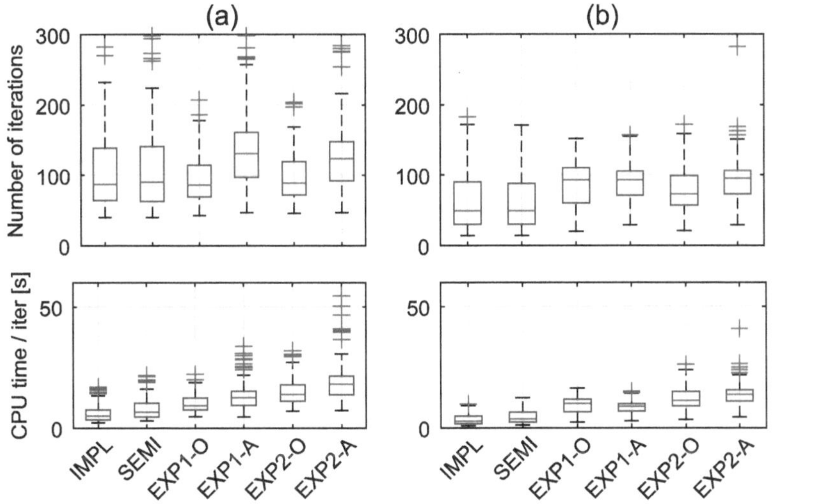

Fig. 3.10 Number of iterations and total CPU time to convergence represented in the form of a box plot for the two sets of solutions. (**a**) First cost function without smoothing: $k_1 = 1e0$, $k_2 = 0$. (**b**) Second cost function with smoothing: $k_1 = 1e0$, $k_2 = 1e5$

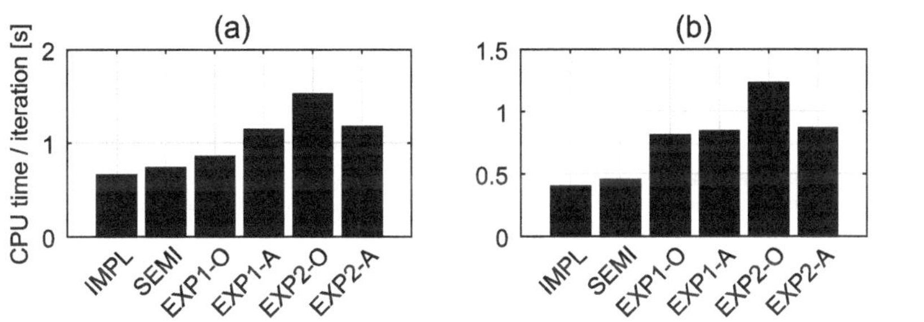

Fig. 3.11 Mean CPU time per iteration for the two sets of solutions. (**a**) First cost function without smoothing: $k_1 = 1e0$, $k_2 = 0$. (**b**) Second cost function with smoothing: $k_1 = 1e0$, $k_2 = 1e5$

forms exhibit similar performances in terms of CPU time when formulated as ODE and DAE systems (see Fig. 3.10). However, the Order-N implementation yields a lower CPU time per iteration when formulated as a DAE system (see Fig. 3.11). Its poor performances in terms of total CPU time are explained by its high number of iterations to reach convergence.

3.4.3.3 Discussion

In this section, we aim at giving some insights about the obtained results. From the simulations, it is clear that the implicit and the semi-explicit formulations perform better than the explicit ones. From a numerical point of view, the DAE form leads

Table 3.3 Size of the NLP associated to the OCP for the second cost definition (smoothed actuation)

Number of	IMP	SEMI	EXP1-O	EXP1-A	EXP2-O	EXP2-A
Variables	8812	8812	6412	8812	6412	8812
Constraint equations	8427	8427	6027	8427	6027	8427
nzt[a] in constraint Jacobian	72,435	72,435	62,519	69,635	62,519	69,635
nzt in Lagrangian Hessian	39,021	39,021	35,286	35,021	35,286	35,021

[a]nzt stands for non-zero terms

Table 3.4 Dynamic formulations and their generation for the studied MBS

	Algorithm	Dynamic terms	# Flops	# Eqs
IMPL	Original NER	$\mathbf{\Phi}(\ddot{\mathbf{q}}, \dot{\mathbf{q}}, \mathbf{q})$	460	114
SEMI	Modified NER	$\mathbf{M}(\mathbf{q})$ and $\mathbf{c}(\dot{\mathbf{q}}, \mathbf{q})$	830	264
EXP1	Modified NER + Cholesky	$\gamma(\dot{\mathbf{q}}, \mathbf{q}, \mathbf{Q})$	991	328
EXP2	Order-N	$\nu(\dot{\mathbf{q}}, \mathbf{q}, \mathbf{Q})$	1784	395

to a larger problem as both the number of equations and variables is higher (see Table 3.3). This difference results from the treatment of joint accelerations as algebraic variables which is mandatory for the implicit and semi-explicit dynamics. Nevertheless, compared to a pure ODE representation, the differential equation $\dot{\mathbf{x}} = \mathbf{f}(\mathbf{x}, \mathbf{u}, \mathbf{z})$ is trivial in the DAE case. In addition, the DAE formulation allows the definition of the terminal equilibrium constraint directly in terms of the algebraic variables, which can be advantageous as the expression of the equilibrium constraints is linear in that case.

The explanation for the performance differences lies in the derivation of the dynamics itself. The different formulations are all written in the form of a function generated symbolically with Robotran. Because they inherit different complexity, the number of equations and floating point operations associated with each formulation differ as shown in Table 3.4. The $\mathcal{O}(N)$ computation of the $\mathbf{\Phi}$ term does in fact lead to a lower number of operations, about half of the $\mathbf{M}$ and $\mathbf{c}$ terms of the semi-explicit dynamics for the system at hand. Regarding the explicit formulations, the NER modified algorithm with Cholesky decomposition leads to a number of flops comparable with the semi-explicit form. This is because the considered MBS is not too large and the Cholesky decomposition is not critical for the number of joints (i.e., 6 for the benchmark). However, the Order-N method yields a high number of operations, that is more than 4 times the number of operations for the implicit form. It may seem strange as the number of operations for the Order-N method is supposed to increase linearly with the size of the system. The problem is that this linear behavior becomes only interesting for a greater number of joints in the MBS (i.e., around 15 according to [8]).

While the explicit expressions clearly differ as they are derived differently, the implicit and the semi-explicit equations are exactly the same but expressed in different forms (see Sects. 3.3.2.1 and 3.3.2.2). As these equations are identical, their derivative should also be the same. As a matter of fact, the implicit and semi-explicit formulations lead to the exact same number of non-zero terms in the expression of the Jacobian and the Hessian for the NLP, which is already a good indicator (see Table 3.3). The fact that the equations are identical should lead to the same iteration path while solving the numerical OCP. This is observed in the case of the smoothed actuation (see Fig. 3.10b) with an almost exact same number of iterations in the 231 tasks. For the case of non-smoothed actuation (see Fig. 3.10a), a similar but non-identical number of iterations is observed for the two formulations. These differences are attributed to the high numerical sensitivity of this Optimal Control Problem. This could cause the NLP solver to use different strategies for the two dynamic formulations. Finally, the implicit and semi-explicit formulations are more frequently trapped in higher local minima than the explicit ones. Some additional investigations would be necessary to explain this observation.

We are aware that treating the explicit dynamics with a collocation scheme may not be the best choice and that a multiple shooting could be more appropriated [6]. As a matter of fact, multiple shooting for the explicit formulations have also been investigated in this work but was abandoned as it led to lower performances in terms of robustness (i.e., lower success rate for the extreme point and slower convergence). The reason could be the high dynamics of the system and the strong terminal equilibrium constraint. In order to verify this assumption, a simpler multibody system could be studied with a smoother formulation of the OCP (i.e., with no terminal constraint).

3.5 Conclusion

This paper analyzes different formulations of the multibody dynamics and compare their performances in terms of convergence, accuracy and computation cost in the context of Optimal Control. The aim is to provide some insights on existing MBS formulations and their relevance applicability in the formulation and solution of Optimal Control Problems. Among the different modeling techniques, we focus on the derivation of the equations, using a relative coordinates approach, with the Newton-Euler Recursive scheme and the Order-N algorithm.

First, a framework for the formulation of Optimal Control Problem has been introduced. It was based on a semi-explicit DAE representation of the dynamics well suited for the treatment of multibody systems. Other formulations could have been considered (e.g., full implicit) but would have required specific integrators (i.e., DAE integrators). Regarding the discretization of the OCP, the direct collocation scheme is selected here. This decision is taken following the considerations found in [6], but also because multiple shooting methods are less robust. Then, the different formulations for the multibody dynamics are introduced and their derivations are discussed. From the Newton Euler recursive scheme, three formulations are derived: the implicit, the semi-explicit and an explicit one. Then, a second explicit formulation is derived from the Order-N algorithm.

Finally, a case study is presented in a benchmark approach for the unconstrained MBS Optimal Control. The system consists in a 3D robot arm with an inverted pendulum attached to the end effector. An OCP is formulated with the goal to reach a point in the 3D space for the end effector after a preset time of 2 s. At the end, the system has to be at equilibrium with zero acceleration and the actuation is minimized on the overall time interval. This task is realized for a grid of target points in the 3D space and for two different definitions of the cost function resulting in two sets of tasks.

Results showed that the implicit and semi-explicit formulations are generally performing better in terms of CPU time and number of iterations. This can be explained by a more compact formulation of the dynamics which leads to a lower CPU cost per iteration. However, these two formulations appear to be more sensitive to higher local minima when the dynamics is stronger (i.e., in the case of non-smoothed actuation). This could be a drawback in applications such as offline trajectory planning where the quality of the solution is more important than its computation cost. On the other hand, the explicit formulations show larger CPU time to convergence but ensure a higher quality of the solution, as a lesser number of high local minima are observed. It seems that the Order-N explicit form works better with a DAE representation, while a pure ODE seems more appropriate for the NER explicit form. For the first set of solutions, with non-smoothed actuation, the differences in the number of iterations for the implicit and semi-explicit formulations are attributed to the strong numerical sensitivity of the problem. More investigation should be carried out on that subject. Nevertheless, this study shows the importance of the dynamic formulations on the OCP solving performances for unconstrained MBS.

There are many perspectives to extend this work. First, it would be appropriate to carry out additional simulations with different OCP formulations and different multibody systems in order to verify the observed tendencies. It has been shown that the way the dynamics is derived has a huge impact on the solver for the OCP as it strongly influences the CPU time per iteration. A recent release of Robotran offers symbolic tools for obtaining a fully analytical sensitivity of MBS equations based on the recursive differentiation of the multibody formulation. It will be interesting to see whether this could lead to better performances than when using the algorithmic differentiation provided by an optimization software like CasADi.

Acknowledgements We would like to thank Dr. Joris Gillis from KUL for his help in the formulation and solving of the Optimal Control Problems with CasADi. Quentin Docquier is a FRIA Grant Holder of the Fonds de la Recherche Scientifique – FNRS, Belgium.

References

1. Andersson, J., Gillis, J., Horn, G., Rawings, J.B., Diehl, M.: CasADi – software framework for nonlinear optimization and optimal control. In: Mathematical Programming Computation (2018)
2. Bastos, G., Seifried, R., Bruls, O.: Inverse dynamics of serial and parallel underactuated multibody system using a DAE optimal control approach. Multibody Sys. Dyn. **30**, 359–376 (2013)

3. Baumgarte, J.: Stabilization of constraints and integrals of motion. Comput. Methods Appl. Mech. Eng. **1**, 1–16 (1972)
4. Betts, J.T.: Practical Methods for Optimal Control and Estimation Using Nonlinear Programming, 2nd edn. SIAM, Philadelphia (2010)
5. Björkenstam, S., Carlson, J.S., Lennartson, B.: Exploiting sparsity in the discrete mechanics and optimal control method with application to human motion planning. In: Proceedings of 11th IEEE International Conference on Automation Science and Engineering, Gothenburg, pp. 769–774 (2015)
6. Diehl, M., Bock, H., Diedam, H., Wieber, P.B.: Fast direct multiple shooting algorithms for optimal robot control. In: Diehl M., Mombaur, K. (eds.) Fast Motions in Biomechanics and Robotics. Lecture Notes in Control and Information Sciences, vol. 340. Springer, Berlin/New York (2006)
7. Docquier, N., Poncelet, A., Fisette, P.: Robotran: a powerful symbolic generator of multibody models. Mech. Sci. **4**, 199–219 (2013)
8. Fisette, P.: Génération Symbolique des Equations du Mouvement de Systèmes Multicorps et Application dans le Domaine Ferroviaire. Ph.D. thesis, Université catholique de Louvain, Louvain-la-Neuve, Belgium (1994)
9. Fisette, P., Samin, J.C.: Symbolic generation of large multibody system dynamic equations using a new semi-explicit Newton/Euler recursive scheme. Arch. Appl. Mech. **66**, 187–199 (1996)
10. Griewank, A., Walther, A.: Evaluating Derivatives: Principles and Techniques of Algorithmic Differentiation. Other Titles in Applied Mathematics, vol. 105. SIAM, Philadelphia (2008)
11. Gros, S., Zanon, M., Diehl, M.: Nonlinear MPC and MHE for mechanical multi-body systems with application to fast tethered airplanes. In: IFAC Conference on Nonlinear Model Predictive Control, Noordwijkerhout, pp. 86–93 (2012)
12. Iwamura, M., Eberhard, P., Schiehlen W., Seifried R.: A general purpose algorithm for optimal trajectory planning of closed loop multibody systems. In: Arczewski, K. (ed.) Multibody Dynamics: Computational Methods and Applications. Computational Methods in Applied Sciences, vol. 23. Springer, Dordrecht (2011)
13. Luh, J.-Y.-S., Walker, N.-W., Paul, R.-P.-C.: On-line computational scheme for mechanical manipulators. J. Dyn. Syst. Meas. Control **102**, 69–76 (1980)
14. Manara, S., Gabiccini, M., Artoni, A., Diehl, M.: On the integration of singularity-free representations of SO(3) for direct optimal control. Nonlinear Dyn. **90**, 1223–1241 (2017)
15. Samin, J.C., Fisette, P.: Symbolic Modeling of Multibody Systems. Kluwer, Dordrecht (2003)
16. Schwertassek, R., Rulka, W.: Aspects of efficient and reliable multibody systems simulation. In: Haug, E.J., Deyo, R.C. (eds.) Real-Time Integration Methods for Mechanical System Simulation. NATO ASI Series F, vol. 69, pp. 55–96. Springer, Berlin/Heidelberg (1989)
17. Siebert, R., Betsch, P.: A Hamiltonian conserving indirect optimal control method for multibody dynamics. In: Proceedings in Applied Mathematics and Mechanics, vol. 12. Wiley-VCH Verlag GmbH & Co. KGaA, Weinheim (2012)
18. Verscheure, D., Demeulenaere, B., Swevers, J., De Schutter, J., Diehl M.: Time-energy optimal path tracking for robots: a numerically efficient optimization approach. In: 10th IEEE International Workshop on Advanced Motion Control, Trento, pp. 727–732 (2008)
19. von Stryk, O., Schlemmer, M.: Optimal control of the industrial robot manutec R3. In: Bulirsch, R. (ed.) Computational Optimal Control. ISNM International Series of Numerical Mathematics, vol. 115. Birkhäuser, Basel/Boston (1994)
20. Wächter, A.: An interior point algorithm for large-scale nonlinear optimization with applications in process engineering, Ph.D. thesis, Carnegie Mellon University (2002)
21. Wehage, R.-A., Haug, E. J.: Generalized coordinate partitioning for dimension reduction in analysis of constrained dynamic systems. J. Mech. Des. **134**, 247–255 (1982)

Chapter 4
A Recursive Driving Constraint Approach for Inverse Dynamics Modeling of Flexible Multibody Systems

Saeed Ebrahimi and Arman Mardani

Abstract In this chapter, a novel procedure for inverse dynamics modeling of flexible multibody systems using a driving constraints approach in the form of a forward dynamics analysis is presented. The flexible 3RPR parallel manipulator is chosen here as a flexible multibody system with closed kinematic chains to introduce this approach. The equations of motion are derived using the floating frame of reference formulation. Assuming a prescribed trajectory of the end-effector, the generalized coordinates of the rigid manipulator associated with the prismatic and revolute joints are obtained from an inverse kinematics analysis of the manipulator. This solution is further exploited in a forward dynamics analysis of the flexible manipulator to form the required driving constraints for obtaining the approximate values of the actuating forces and torques. Finally, the inverse dynamics analysis of the flexible manipulator is carried out by including some additional driving constraints associated with the generalized elastic coordinates to obtain the high-accuracy approximate values of the actuating forces and torques for tracking the prescribed trajectory by the end-effector. To numerically validate the approach, the obtained actuating forces and torques are applied to the simulated flexible manipulator to check the final trajectory of the end-effector. The results confirm the accuracy of the proposed approach.

Keywords Flexible parallel robots · 3RPR platform · Floating frame formulation · Inverse dynamics

---

S. Ebrahimi (✉) · A. Mardani
Mechanical Engineering Department, Yazd University, Yazd, Iran
e-mail: ebrahimi@yazd.ac.ir

E. Zahariev, J. Cuadrado (eds.), *IUTAM Symposium on Intelligent Multibody Systems – Dynamics, Control, Simulation*, IUTAM Bookseries 33,
https://doi.org/10.1007/978-3-030-00527-6_4

4.1 Introduction

Inverse dynamics modeling of flexible multibody systems has been usually considered as a challenging task. The complexity involved in this problem is usually due to the existent of the governing dynamic equations of motion fewer than the unknown parameters. Indeed, flexible coordinates yield an undetermined system of equations. In this context, dynamics analysis of the flexible manipulators which are usually divided into serial and parallel platforms can be considered as an important applied problem [1]. The open-loop platform of serial manipulators induces undesired fluctuations of the end-effector due to the flexibility involved [2]. Rodriguez proposed an spatial operator approach to flexible manipulator inverse and forward dynamics by using spatially recursive filtering and smoothing techniques to reduce fluctuations [3]. Ledesma et al. investigated inverse dynamics of spatial open-chain flexible manipulators with lumped and distributed actuators [4]. Mardani and Ebrahimi studied the effect of a sequential PID controller for a long-thin flexible manipulator with low-weight arms in order to reduce the fluctuations of links and end-effector [5]. As another mathematical efforts, the researches proposed in [6, 7] dealt with efficient computation of the inverse dynamics of flexible manipulators. The neural network can be obviously effective to solve the problematic system of equations. Accordingly, the issue of inverse and forward dynamics of flexible serial manipulators has been addressed by using neural networks as an advanced implementation of the artificial intelligence in robot dynamics [8–10]. As an advanced application, the flexible serial manipulators implemented in cooperation activities, are obviously more complicated than the simple serial manipulation missions. An inverse dynamics algorithm for multiple flexible-link manipulators was presented in [11] by Sun et al. to treat such problems. Undesired effects of the flexibility on the path tracking process of the manipulators strongly required advanced control concepts. A global Lagrangian approach is employed in [12] to formulate the system equations of motion, and an iterative procedure is proposed to achieve the end-point trajectory tracking in three-dimensional workspace. Consequently, as a solution, the inverse dynamics of such flexible 3D multibody system was solved using an optimal control method [13]. These investigations were usually conducted for serial manipulators subject to the light payloads. On the other hand, large payloads strongly change the accuracy of approximation of the inverse dynamics and motion analysis due to the flexibility effects [14]. Analysis of the flexible manipulators in the presence of Coulomb friction yields conditional stick-slip dynamics as considered by Trautt and Bayo [15]. A general technique for solving the inverse dynamics of flexible robots has been proposed by Carrera and Serna in 1996 [16].

Parallel robots are composed of kinematic chains with more complicated structure than serial manipulators. The virtual work principle was implemented to analyze direct and inverse dynamics of parallel robots such as 3PRR and 3RPS platforms [17–19]. Plitea et al. investigated inverse dynamics of a rigid five-DOF parallel robot by using the virtual work principle [20]. Regarding natural orthogonal complement, Kordjazi and Akbarzadeh investigated inverse dynamics of a 3-PRR planar parallel manipulator [21]. Asada has shown that dynamic deformations of flexible arms can be represented with use of the virtual coordinate systems which eliminates some terms of the equations of motion and significantly reduces complexity in the inverse dynamics [22]. In dynamic modeling and inverse dynamic analysis of flexible parallel robots, the relation between elastic displacements has been analyzed in [23] to reduce the complexity of the problem regarding the coupling effects of elastic and rigid motions. Furthermore, some researches considered a finite element framework to include flexibility of links for inverse dynamics of vibrating robots [24]. The simplest assumption is to only include one rotational flexible degree for planar links which yields some neglected elastic modes especially in the presence of prismatic joints [25, 26].

As the current research contributes, inclusion of driving kinematic constraints yields a new approach for inverse dynamics modeling of flexible multibody systems such as the flexible 3RPR manipulator which finally includes three rigid and six flexible DOFs. The objective is to find the unknowns which include the motor torques or actuator forces. The forward dynamics of this manipulator has been previously formulated in [26, 27]. Here, a novel procedure for inverse dynamics modeling of this manipulator using a driving constraints approach in the form of forward dynamics problem is presented. The equations of motion are derived using the floating frame of reference formulation [28]. In following, the required steps for implementing this procedure are explained.

4.2 Modeling of Flexible Bodies Using the Floating Frame Formulation

In the context of multibody dynamics, the floating frame formulation is efficiently implemented in order to analyze flexible bodies and derive equations of motion. According to this formulation, the position of a planar flexible body i is defined by means of rigid coordinates $\mathbf{q}_r^i = \begin{bmatrix} \mathbf{R}^{i^{\mathrm{T}}} & \theta^i \end{bmatrix}^{\mathrm{T}}$ and flexible coordinates $\mathbf{q}_f^i$. The position of a typical point p shown in Fig. 4.1 can be defined in the following manner [28].

$$\mathbf{r}_p^i = \mathbf{R}^i + \mathbf{A}^i \left(\overline{\mathbf{u}}_o^i + \mathbf{S}^i \mathbf{q}_f^i \right) \tag{4.1}$$

The vector $\mathbf{R}^i$, matrix $\mathbf{A}^i$ and vector $\overline{\mathbf{u}}_o^i$ respectively denote the position of the coordinate system attached to the body i, the rotation matrix and the coordinates of point p in the undeformed configuration. Moreover, the shape matrix $\mathbf{S}^i$ depends on the conditions and the reference coordinate system of the body. In this contribution, dynamics analysis of the constrained system is investigated using the Eeuler-Lagrange approach whereby the kinematic constraints are derived by including a set of generalized coordinates. Consequently, implementing Lagrange multipliers

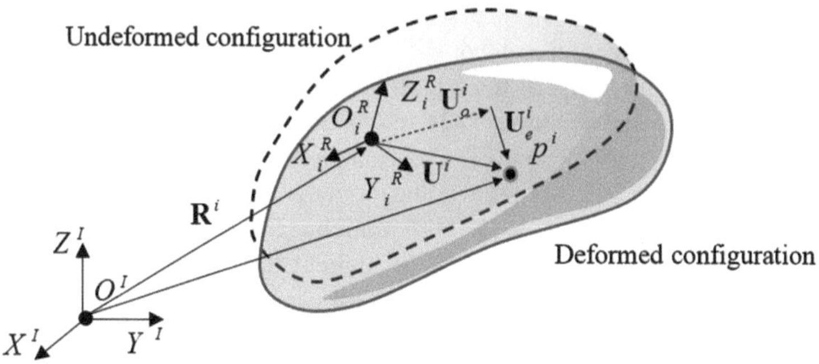

Fig. 4.1 Kinematics of the flexible body

vector, kinematic constraints are augmented by the dynamic equations. According to this approach, the constraint forces appear in the final form of the differential equations which yields a system of coupled nonlinear equations written as

$$\mathrm{M}^i\ddot{\mathrm{q}}^i + \mathrm{K}^i\mathrm{q}^i + \mathrm{C}_{\mathrm{q}^i}^{\mathrm{T}}\lambda = \mathrm{Q}_e^i + \mathrm{Q}_v^i, \quad i = 1, 2, \ldots, n_b \tag{4.2}$$

The generalize coordinates $\mathbf{q}^i = \left[\mathbf{q}_r^{i\mathrm{T}}\ \mathbf{q}_f^{i\mathrm{T}}\right]^{\mathrm{T}}$ includes the rigid and flexible coordinates. The parameter n_b, matrices $\mathbf{M}^i$, $\mathbf{K}^i$, $\mathbf{C}_{\mathrm{q}^i}$ and vector $\boldsymbol{\lambda}^i$ respectively denote number of bodies, symmetric mass matrix, stiffness matrix, Jacobian matrix of the kinematic constraints and the Lagrange multipliers vector. Moreover, the external and centrifugal-Coriolis forces vectors are respectively defined by $\mathbf{Q}_e^i$ and $\mathbf{Q}_v^i$ [28]. The final matrix form of the equations can now be expressed as

$$\begin{bmatrix} \mathrm{m}_{RR}^i & \mathrm{m}_{R\theta}^i & \mathrm{m}_{Rf}^i \\ & m_{\theta\theta}^i & \mathrm{m}_{\theta f}^i \\ \mathrm{sym.} & & \mathrm{m}_{ff}^i \end{bmatrix} \begin{bmatrix} \ddot{\mathrm{R}}^i \\ \ddot{\theta}^i \\ \ddot{\mathrm{q}}_f^i \end{bmatrix} + \begin{bmatrix} \mathbf{0} & \mathbf{0} & \mathbf{0} \\ \mathbf{0} & \mathbf{0} & \mathbf{0} \\ \mathbf{0} & \mathbf{0} & \mathrm{K}_{ff}^i \end{bmatrix} \begin{bmatrix} \mathrm{R}^i \\ \theta^i \\ \mathrm{q}_f^i \end{bmatrix} + \begin{bmatrix} \mathrm{C}_{\mathrm{R}^i}^{\mathrm{T}} \\ \mathrm{C}_{\theta^i}^{\mathrm{T}} \\ \mathrm{C}_{\mathrm{q}_f^i}^{\mathrm{T}} \end{bmatrix} \lambda = \begin{bmatrix} (\mathrm{Q}_e^i)_{\mathrm{R}} \\ (\mathrm{Q}_e^i)_{\theta} \\ (\mathrm{Q}_e^i)_f \end{bmatrix} + \begin{bmatrix} (\mathrm{Q}_v^i)_{\mathrm{R}} \\ (\mathrm{Q}_v^i)_{\theta} \\ (\mathrm{Q}_v^i)_f \end{bmatrix} \tag{4.3}$$

where $i = 1, 2, \ldots, n_b$. The components of the generalized external force vector associated with the applied external force $\mathbf{F}^i$ at a certain point of the body is derived as

$$\left(\mathrm{Q}_e^i\right)_{\mathrm{R}} = \mathbf{F}^i, \quad \left(\mathrm{Q}_e^i\right)_{\theta} = \mathbf{B}^{i\mathrm{T}}\mathbf{F}^i, \quad \left(\mathrm{Q}_e^i\right)_f = \mathrm{S}^{i\mathrm{T}}\mathrm{A}^{i\mathrm{T}}\mathbf{F}^i \tag{4.4}$$

where $\mathbf{B}^i$ and $\mathbf{S}^i$ are defined according to the local position of the applied external force. The components of the centrifugal-Coriolis forces vector $(\mathrm{Q}_v^i)_{\mathrm{R}}$ and $(\mathrm{Q}_v^i)_f$

are expressed, e.g., in [28]. Matrices $\mathbf{I}_l^i$, $\overline{\mathbf{S}}^i$ and $\overline{\mathbf{I}}_o^i$ define the inertia shape integrals. The stiffness matrix associated with elastic coordinates of a flexible link i in a multibody system can be given by

$$\mathbf{K}_{ff}^i = \int_{v^i} \left(\mathbf{D}^i\mathbf{S}^i\right)^T \mathbf{E}^i\mathbf{D}^i\mathbf{S}^i dV^i \tag{4.5}$$

where $\mathbf{D}^i$ is a differential operator defining the relation between the strain and deformation vectors [28]. Moreover, $\mathbf{E}^i$ denotes the symmetric matrix of elastic coefficients of the flexible link assuming linear isotropic material.

4.3 Modeling of the Flexible 3RPR Manipulator

4.3.1 Coordinate System Definition

According to Fig. 4.2, this manipulator is composed of seven links including three rigid cylinders A_iB_i, three flexible links B_iC_i $(i = 1, 2, 3)$, and a moving platform $C_1C_2C_3$ which are connected to the base $A_1A_2A_3$ through three closed kinematic chains. For the sake of brevity, the parameters A_1B_1, A_2B_2, A_3B_3, B_1C_1 , B_2C_2 and B_3C_3 are respectively denoted by the superscripts 1,3,5,2,4, and 6 as the number of links. To distinguish the exponentiation operation from superscripts, the exponents

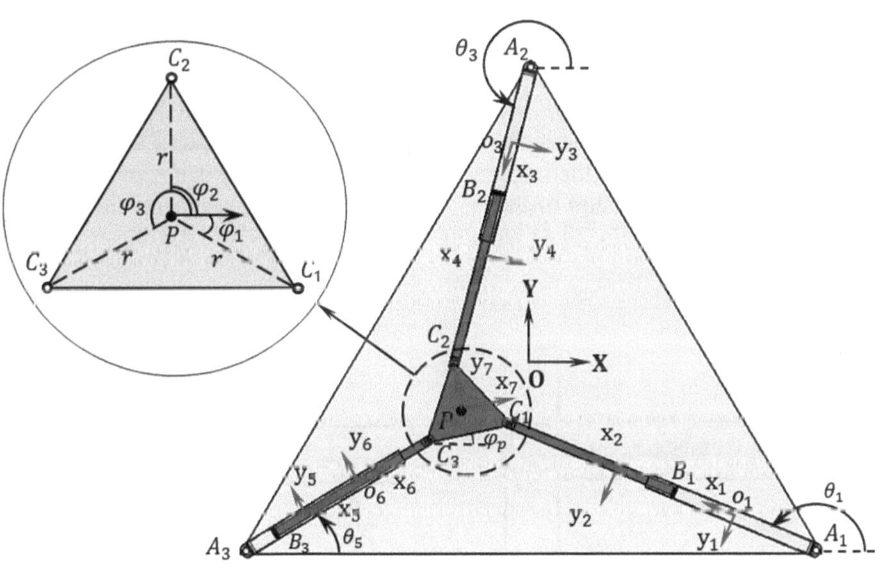

Fig. 4.2 Flexible 3RPR manipulator

appear in parentheses. Three DOFs of the moving platform are determined by the variables x_P, y_P, and ϕ_P. Moreover, a local coordinate system is attached to the center of mass of A_iB_i and another local coordinate system is attached to the point B_i of the link B_iC_i whose positions are defined by the generalized coordinates ($j = 1, \ldots, 6$), $\mathbf{q}_j = [R_{xj}\ R_{yj}\ \ \theta_j]^{\mathrm{T}}$. All the coordinates are represented relative to the fixed frame *XY*. Accordingly, the generalized coordinates vector associated by the rigid motion of the system including 21 variables are represented as

$$\mathbf{q}_r = \left[\mathbf{q}_1^{\mathrm{T}}, \ldots, \mathbf{q}_6^{\mathrm{T}}, x_p y_p\ \phi_p\right]^{\mathrm{T}} = \left[R_{x1}\ R_{y1}\ \theta_1,\ \ldots\ , R_{x6}\ R_{y6}\ \theta_6, x_p y_p\ \phi_p\right]^{\mathrm{T}} \tag{4.6}$$

4.3.2 Flexibility Modeling

In this research, only the intermediate links B_iC_i are assumed to be flexible. The shape matrix of each flexible link is determined by assuming two nodes located at points B_i and C_i, and a local coordinate system at point B_i. Equation (4.7) defines the shape matrix

$$\mathrm{S}_b = \begin{bmatrix} 1-\xi & 0 & 0 & \xi & 0 & 0 \\ 0 & 1-3(\xi)^2-2(\xi)^3 & l\left[\xi-2(\xi)^2+(\xi)^3\right] & 0 & 3(\xi)^2-2(\xi)^3 & l\left[(\xi)^3-(\xi)^2\right] \end{bmatrix} \tag{4.7}$$

where $\xi = (x/l)$. The flexible coordinate vector is written as the following equation

$$\overline{\mathrm{q}}_f = \left[\overline{q}_{f1}, K, \overline{q}_{f6}\right]^{\mathrm{T}} \tag{4.8}$$

wherein the variables $\overline{q}_{f1}$ and $\overline{q}_{f4}$ are respectively the longitudinal displacement of the start and end points. Furthermore, $\overline{q}_{f2}$ and $\overline{q}_{f5}$ represent the lateral displacement of these points. The slop at the start-end points are respectively denoted by $\overline{q}_{f3}$ and $\overline{q}_{f6}$. Regarding the inclusion of the rigid modes in the displacements field, some flexible coordinates of Eq. (4.8) have to be eliminated. Figure 4.3 illustrates the

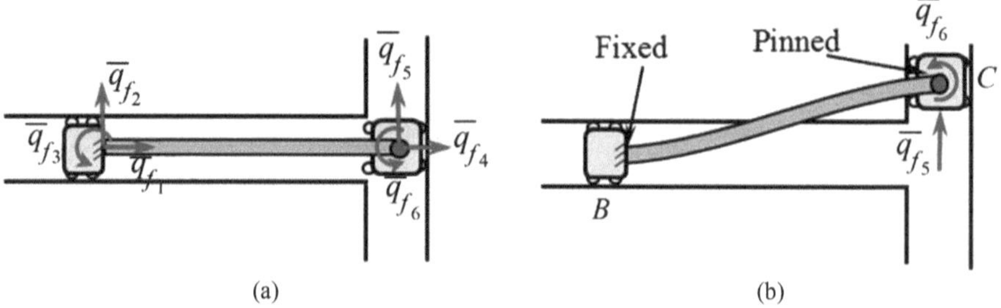

Fig. 4.3 Definition of flexible intermediate links: (**a**) undeformed, (**b**) deformed configuration

deformed and undeformed configurations of the flexible link. The bottom portion of the flexible link is firmly constrained to remain parallel to the rigid guide. The actuated prismatic joints moving the flexible links are composed of cylinder-piston joints which do not allow any slope variation of the bottom portion of the flexible link. According to this figure, the boundary conditions of the flexible link are written as

$$\overline{u}_{f1}(0) = 0, \quad \overline{u}_{f2}(0) = 0, \quad \overline{u}_{f1}(l) = 0 \tag{4.9}$$

Consequently, considering Eqs. (4.1) and (4.9), some coordinates are equal to zero as Eq. (4.10) shows

$$\overline{q}_{f1} = 0, \quad \overline{q}_{f2} = 0, \quad \overline{q}_{f3} = 0, \quad \overline{q}_{f4} = 0 \tag{4.10}$$

Therefore, the flexible coordinates vector is rewritten as

$$\mathbf{q}_f = \left[\overline{q}_{f5}, \overline{q}_{f6}\right]^{\mathrm{T}} \tag{4.11}$$

Equation (4.12) explains the relation between two vectors $\overline{\mathbf{q}}_f$ and $\mathbf{q}_f$

$$\overline{\mathbf{q}}_f = \left[\overline{q}_{f1}, K, \overline{q}_{f6}\right]^{\mathrm{T}} = \begin{bmatrix} 0\,0\,0\,0\,1\,0 \\ 0\,0\,0\,0\,0\,1 \end{bmatrix}^{\mathrm{T}} \begin{bmatrix} \overline{q}_{f5} \\ \overline{q}_{f6} \end{bmatrix} \Rightarrow \overline{\mathbf{q}}_f = \mathbf{B}_r \mathbf{q}_f \tag{4.12}$$

Hence, the displacements field vector $\overline{\mathbf{u}}_f$ is consequently formulized as

$$\overline{\mathrm{u}}_f = \mathrm{S}_b \overline{\mathrm{q}}_f = \mathrm{S}_b \mathrm{B}_r \mathrm{q}_f = \mathrm{S}\,\mathrm{q}_f \tag{4.13}$$

According to Eq. (4.13), the shape matrix is rearranged as

$$\mathrm{S} = \mathrm{S}_b \mathrm{B}_r = \begin{bmatrix} 0 & 0 \\ 3(\xi)^2 - 2(\xi)^3 & l\left[(\xi)^3 - (\xi)^2\right] \end{bmatrix} \tag{4.14}$$

As previous formulations pointed out, two flexible coordinates $q_{f1} = \overline{q}_{f5}$ and $q_{f2} = \overline{q}_{f6}$ are sufficient to explain the flexible displacements. These coordinates respectively define the lateral displacement and the slop at the end point of the flexible link. The total flexible coordinate vector for the whole platform can hence be written as

$$\mathbf{q}_f = \left[q_{f1}^2 \; q_{f2}^2 \; q_{f1}^4 \; q_{f2}^4 \; q_{f1}^6 \; q_{f2}^6\right]^{\mathrm{T}} \tag{4.15}$$

As a conclusion, according to the system of the generalized coordinates, the flexible platform includes three rigid and six flexible DOFs.

4.3.3 Kinematic Constraints

Regarding 21 generalized coordinates describing three rigid DOFs, 18 kinematic constraints have to be implemented to mathematically describe the revolute and prismatic joints. It means when each planar revolute or prismatic joint reduces two degrees of freedom, two kinematic constraint equations have to be written. As the first step to derive the dynamics, the nonlinear algebraic equations of the constraints have to be formulized. For the sake of brevity, only the constraint equations of the joints at points A_1, B_1 and, C_1 are described. Two constraints of the revolute joint are written as

$$\begin{bmatrix} R_{x1} \\ R_{y1} \end{bmatrix} + \begin{bmatrix} \cos\theta_1 & -\sin\theta_1 \\ \sin\theta_1 & \cos\theta_1 \end{bmatrix} \begin{bmatrix} -\frac{l_1}{2} \\ 0 \end{bmatrix} = \begin{bmatrix} X_{A1} \\ Y_{A1} \end{bmatrix} \tag{4.16}$$

wherein the parameter l_1 defines the length of the link A_1B_1. Moreover, the coordinates of the point A_1 are denoted by (X_{A1}, Y_{A1}) relative to the XY coordinate system. Consequently, two constraints of the prismatic joint can be described as

$$\begin{cases} \mathbf{h}_2^T \boldsymbol{\rho}_2 = 0 \\ \theta_1 - \theta_2 = 0 \end{cases} \tag{4.17}$$

The vector $\boldsymbol{\rho}_2$ defines the direction of sliding of the prismatic joint, and accordingly, the unit vector $\mathbf{h}_2$ defines the perpendicular direction to $\boldsymbol{\rho}_2$ which both vectors can be expressed as

$$\begin{aligned} \boldsymbol{\rho}_2 &= \mathbf{r}_{B1} - \mathbf{r}_{A1} = \begin{bmatrix} R_{x2} \\ R_{y2} \end{bmatrix} - \begin{bmatrix} R_{x1} \\ R_{y1} \end{bmatrix} - \begin{bmatrix} \cos\theta_1 & -\sin\theta_1 \\ \sin\theta_1 & \cos\theta_1 \end{bmatrix} \begin{bmatrix} -\frac{l_1}{2} \\ 0 \end{bmatrix}, \\ \mathbf{h}_2 &= \begin{bmatrix} \cos\theta_2 & -\sin\theta_2 \\ \sin\theta_2 & \cos\theta_2 \end{bmatrix} \begin{bmatrix} 0 \\ 1 \end{bmatrix} \end{aligned} \tag{4.18}$$

Finally, by setting the coordinates of C_1 belonging to the end-effector equal to the coordinates of the end point of link B_1C_1, two constraints of the revolute joint at C_1 can be derived as

$$\begin{aligned} &\begin{bmatrix} R_{x2} \\ R_{y2} \end{bmatrix} + \begin{bmatrix} \cos\theta_2 & -\sin\theta_2 \\ \sin\theta_2 & \cos\theta_2 \end{bmatrix} \begin{bmatrix} l_2 \\ q_{f1}^2 \end{bmatrix} - \begin{bmatrix} x_p \\ y_p \end{bmatrix} \\ &\quad - \begin{bmatrix} \cos\phi_p & -\sin\phi_p \\ \sin\phi_p & \cos\phi_p \end{bmatrix} \begin{bmatrix} r\cos\varphi_1 \\ r\sin\varphi_1 \end{bmatrix} = \begin{bmatrix} 0 \\ 0 \end{bmatrix} \end{aligned} \tag{4.19}$$

where, according to Fig. 4.2, the parameter r and the variable φ_1 respectively define the distance between points p and C_1, and the angle between pC_1 and x_7.

4.4 Derivation of the Equations of Motion

4.4.1 Mass Matrix of the Flexible Links

In this section, the mass matrix of the flexible link B_1C_1 using the relations of Sect. 2 is derived. First, the inertia shape integral can be written as

$$I_l^2 = \int_0^{l_2} \rho \begin{bmatrix} x \\ 0 \end{bmatrix} a\, dx = \begin{bmatrix} \frac{m_2 l_2}{2} & 0 \end{bmatrix}^{\mathrm{T}} \tag{4.20}$$

where a denotes the area of the link's cross section. Then, the matrix $\overline{S}^i$ can be given by

$$\overline{S}^2 = \int_0^{l_2} \rho a S\, dx = \int_0^1 \rho a l_2 S\, d\xi = m_2 \int_0^1 \begin{bmatrix} 0 & 0 \\ 3(\xi)^2 - 2(\xi)^3 & l_2\left[(\xi)^3 - (\xi)^2\right] \end{bmatrix} d\xi = \frac{m_2}{12}\begin{bmatrix} 0 & 0 \\ 6 & -l_2 \end{bmatrix} \tag{4.21}$$

The components of mass matrix can be obtained using the formulations presented in [28]. Therefore, the matrices m_{RR}^2, $m_{R\theta}^2$ and m_{Rf}^2 are defined as

$$m_{RR}^2 = \begin{bmatrix} m_2 & 0 \\ 0 & m_2 \end{bmatrix} \tag{4.22}$$

$$m_{R\theta}^2 = \int_{V^2} \rho^2 B^2 dV^2 = A_\theta \left(I_l^2 + S^2 q_f^2 \right) = \begin{bmatrix} -\sin\theta_2 & -\cos\theta_2 \\ \cos\theta_2 & -\sin\theta_2 \end{bmatrix} \left(\begin{bmatrix} \frac{m_2 l_2}{2} \\ 0 \end{bmatrix} + \frac{m_2}{12} \begin{bmatrix} 0 & 0 \\ 6 & -l_2 \end{bmatrix} \begin{bmatrix} q_{f1}^2 \\ q_{f2}^2 \end{bmatrix} \right) \tag{4.23}$$

$$m_{Rf}^2 = \int_{V^2} \rho^2 A^2 S^2 dV^2 = A^2 \overline{S}^2 = \frac{m_2}{12} \begin{bmatrix} \cos\theta_2 & -\sin\theta_2 \\ \sin\theta_2 & \cos\theta_2 \end{bmatrix} \begin{bmatrix} 0 & 0 \\ 6 & -l_2 \end{bmatrix} = \frac{m_2}{12} \begin{bmatrix} -6\sin\theta_2 & l_2 \sin\theta_2 \\ 6\cos\theta_2 & -l_2\cos\theta_2 \end{bmatrix} \tag{4.24}$$

The submatrix $m_{\theta\theta}^i$ of the mass matrix can now be written as the summation of three components as

$$m_{\theta\theta}^2 = \int_{V^2} \rho^2 B^{2\mathrm{T}} B^2 dV^2 = \left(m_{\theta\theta}^2\right)_{rr} + \left(m_{\theta\theta}^2\right)_{rf} + \left(m_{\theta\theta}^2\right)_{ff} \tag{4.25}$$

The components $\left(m_{\theta\theta}^2\right)_{rr}$, $\left(m_{\theta\theta}^2\right)_{rf}$ and $\left(m_{\theta\theta}^2\right)_{ff}$ can be extracted from Eq. (4.25) and written as

$$\left(m_{\theta\theta}^2\right)_{rr} = \int_0^{l_2} \rho_2 a \left[x \; 0\right] \begin{bmatrix} x \\ 0 \end{bmatrix} dx = \frac{m_2 (l_2)^2}{3} \tag{4.26}$$

$$\left(m_{\theta\theta}^2\right)_{rf} = 2\int_{\mathbf{0}}^{1} m_2 \left[l_2\xi \; 0\right] \begin{bmatrix} 0 & 0 \\ 3(\xi)^2 - 2(\xi)^3 & l_2\left[(\xi)^3 - (\xi)^2\right] \end{bmatrix} \begin{bmatrix} q_{f1}^2 \\ q_{f2}^2 \end{bmatrix} d\xi = 0 \tag{4.27}$$

$$\begin{aligned}\left(m_{\theta\theta}^2\right)_{ff} = \int_0^1 m_2 \left[q_{f1}^2 \; q_{f2}^2\right] \begin{bmatrix} 0 & 3(\xi)^2 - 2(\xi)^3 \\ 0 & l_2\left[(\xi)^3 - (\xi)^2\right] \end{bmatrix} \\ \begin{bmatrix} 0 & 0 \\ 3(\xi)^2 - 2(\xi)^3 & l_2\left[(\xi)^3 - (\xi)^2\right] \end{bmatrix} \begin{bmatrix} q_{f1}^2 \\ q_{f2}^2 \end{bmatrix} d\xi = \\ m_2 \left[\frac{13}{35}\left(q_{f1}^2\right)^2 - \frac{11\left(q_{f1}^2 q_{f2}^2\right) l_2}{105} + \frac{\left(q_{f2}^2\right)^2 (l_2)^2}{105}\right]\end{aligned} \tag{4.28}$$

The submatrix $m_{\theta\theta}^i$ can now be obtained using Eqs. (4.25, 4.26, 4.27 and 4.28) as

$$m_{\theta\theta}^2 = m_2 \left[\frac{(l_2)^2}{3} + \frac{13}{35}\left(q_{f1}^2\right)^2 - \frac{11\left(q_{f1}^2 q_{f2}^2\right) l_2}{105} + \frac{\left(q_{f2}^2\right)^2 (l_2)^2}{105}\right] \tag{4.29}$$

The submatrix $\mathrm{m}_{\theta f}^2$ is then derived as

$$\begin{aligned}\mathrm{m}_{\theta f}^2 = \int_{V^2} \rho^2 \mathrm{B}^{2\mathrm{T}} \mathrm{A}^2 \mathrm{S}^2 dV^2 = \int_0^1 \rho_2 a (l_2)^2 \left[\xi \; 0\right] \begin{bmatrix} 0 & 1 \\ -1 & 0 \end{bmatrix} \\ \begin{bmatrix} 0 & 0 \\ 3(\xi)^2 - 2(\xi)^3 & l_2\left[(\xi)^3 - (\xi)^2\right] \end{bmatrix} d\xi = m_2 l_2 \left[\tfrac{7}{20} \; -\tfrac{l_2}{20}\right]\end{aligned} \tag{4.30}$$

Finally, the submatrix m_{ff}^2 can be given by

$$\begin{aligned}\mathrm{m}_{ff}^2 = \int_{V^2} \rho^2 \mathrm{S}^{2\mathrm{T}} \mathrm{S}^2 dV^2 = \int_{\mathbf{0}}^1 m_2 \begin{bmatrix} 0 & 3(\xi)^2 - 2(\xi)^3 \\ 0 & l_2\left[(\xi)^3 - (\xi)^2\right] \end{bmatrix} \\ \begin{bmatrix} 0 & 0 \\ 3(\xi)^2 - 2(\xi)^3 & l_2\left[(\xi)^3 - (\xi)^2\right] \end{bmatrix} d\xi = m_2 \begin{bmatrix} \frac{13}{35} & -\frac{11\, l_2}{210} \\ -\frac{11 l_2}{210} & \frac{(l_2)^2}{210} \end{bmatrix}\end{aligned} \tag{4.31}$$

Substituting the submatrices from Eqs. (4.22, 4.23 and 4.24) and (4.29, 4.30, and 4.31) into Eq. (4.3) yields a 5 × 5 mass matrix for the flexible link B_1C_1. The mass matrix of other flexible links can be derived in a similar manner. In addition, the 3×3 mass matrix of each rigid link can be obtained simply.

4.4.2 Stiffness Matrix of the Flexible Links

The stiffness matrix of the flexible link B_1C_1 can be written as

$$\mathrm{K}_{ff} = \int_0^{l_2} \begin{bmatrix} 0 & \frac{6}{(l_2)^2} - \frac{12x}{l_2} \\ 0 & l_2\left(\frac{6x}{(l_2)^3} - \frac{2}{(l_2)^2}\right) \end{bmatrix} \begin{bmatrix} Ea & 0 \\ 0 & EI \end{bmatrix} \begin{bmatrix} 0 & 0 \\ \frac{6}{(l_2)^2} - \frac{12x}{l_2} & l_2\left(\frac{6x}{(l_2)^3} - \frac{2}{(l_2)^2}\right) \end{bmatrix} dx = EI \begin{bmatrix} \frac{12}{(l_2)^3} & -\frac{6}{(l_2)^2} \\ -\frac{6}{(l_2)^2} & \frac{4}{l_2} \end{bmatrix} \tag{4.32}$$

where I and E denote respectively the area moment of inertia and Young's modulus of B_1C_1.

4.4.3 Coriolis and Centrifugal Forces

The components of the Coriolis and centrifugal forces can be obtained by substituting the required parameters and variables into their associated relations in [28]. The matrices $\bar{\mathrm{I}}_o^2$ and $\tilde{\mathrm{S}}^2$ contain zero values for the flexible link B_1C_1. Hence, one can write

$$\begin{aligned}(\mathrm{Q}_v^2)_R = (\dot{\theta}_2)^2 \begin{bmatrix} \cos\theta_2 & -\sin\theta_2 \\ \sin\theta_2 & \cos\theta_2 \end{bmatrix} \left(\frac{m_2}{12} \begin{bmatrix} 0 \\ 6q_{f1}^2 - l_2 q_{f2}^2 \end{bmatrix} + \begin{bmatrix} \frac{m_2 l_2}{2} \\ 0 \end{bmatrix} \right) \\ - \frac{m_2\dot{\theta}_2}{6} \begin{bmatrix} -\sin\theta_2 & -\cos\theta_2 \\ \cos\theta_2 & -\sin\theta_2 \end{bmatrix} \begin{bmatrix} 0 \\ 6\dot{q}_{f1}^2 - l_2\dot{q}_{f2}^2 \end{bmatrix}\end{aligned} \tag{4.33}$$

$$\begin{aligned}(\mathrm{Q}_v^2)_\theta &= -2\dot{\theta}_2 m_2 \begin{bmatrix} \dot{q}_{f1}^2 & \dot{q}_{f2}^2 \end{bmatrix} \begin{bmatrix} \frac{13}{35} & -\frac{11\,l_2}{210} \\ -\frac{11 l_2}{210} & \frac{(l_2)^2}{210} \end{bmatrix} \begin{bmatrix} q_{f1}^2 \\ q_{f2}^2 \end{bmatrix} \\ &= -2\dot{\theta}_2 m_2 \left(\left(\tfrac{13}{35}\dot{q}_{f1}^2 - \tfrac{11\,l_2}{210}\dot{q}_{f2}^2 \right) q_{f1}^2 + \left(-\tfrac{11\,l_2}{210}\dot{q}_{f1}^2 + \tfrac{l_2}{105}\dot{q}_{f2}^2 \right) q_{f2}^2 \right)\end{aligned} \tag{4.34}$$

$$\left(\mathrm{Q}_v^2\right)_f = (\dot{\theta}_2)^2 m_2 \begin{bmatrix} \frac{13}{35} & -\frac{11\,l_2}{210} \\ -\frac{11 l_2}{210} & \frac{(l_2)^2}{210} \end{bmatrix} \begin{bmatrix} q_{f1}^2 \\ q_{f2}^2 \end{bmatrix} = (\dot{\theta}_2)^2 m_2 \begin{bmatrix} \frac{13}{35} q_{f1}^2 - \frac{11\,l_2}{210} q_{f2}^2 \\ -\frac{11 l_2}{210} q_{f1}^2 + \frac{(l_2)^2}{210} q_{f2}^2 \end{bmatrix} \tag{4.35}$$

4.4.4 External Forces

Considering that the robot is planar, the gravitational forces of all links are perpendicular to the plane of motion and consequently, share no effect to the external forces vector. In addition, the effect of kinematic constraint forces has already been considered though the Lagrange multipliers vector and therefore, they are not supposed as external forces. It remains only the actuating forces or torques moving the manipulator. Two following cases are hence considered to formulate the external forces. First, the external torques applied to the revolute joints A_i of the rigid links A_iB_i, and the external forces applied to the prismatic joints B_i of the flexible links B_iC_i, see Fig. 4.2.

4.4.4.1 Case 1: Motor Torques

In this case, wherein the motor torques are directly applied to the rigid links A_iB_i, only the external forces associated with these links contain non-zero values. For example, assuming that the torque T_1 is applied to the joint A_1 of the link A_1B_1, one can wirte

$$\left(\mathrm{Q}_e^1\right)_R = [0\,0]^{\mathrm{T}}, \quad \left(\mathrm{Q}_e^1\right)_\theta = T^1 \tag{4.36}$$

The generalized external forces of other rigid links can be written in a similar manner, whereas they vanish for the flexible links.

4.4.4.2 Case 2: Linear Actuator Forces

Considering that the linear actuator forces F_a^2 is applied to the flexible link B_1C_1at joint B_1, one can write $\xi = 0$. Therefore, the vector $\mathbf{B}^i$ in Eq. (4.4) and the shape matrix $\mathbf{S}^i$ in Eq. (4.14) contain only zero values which yeilds

$$\begin{aligned} (\mathrm{Q}_e^2)_R &= \left[\, F_a^2 \cos\theta_2 \;\; F_a^2 \sin\theta_2 \,\right]^{\mathrm{T}}, \quad (\mathrm{Q}_e^2)_\theta = 0, \\ (\mathrm{Q}_e^2)_f &= \mathrm{S}^{2\mathrm{T}}\mathrm{A}^{2\mathrm{T}}\mathbf{F}^2 = [0\,0]^{\mathrm{T}} \end{aligned} \tag{4.37}$$

The generalized external forces of other flexible links can be written in a similar manner, whereas they vanish for the rigid links as well as the end-effector.

4.4.5 General Equations of Motion of the Flexible 3RPR Manipulator

The general equation of motion is composed of the equations of motion of four rigid and three flexible links. Consequently, the general equation of motion can be written as, see [28],

$$\begin{bmatrix} \mathrm{M} & \mathrm{C}_{\mathrm{q}}^{\mathrm{T}} \\ \mathrm{C}_{\mathrm{q}} & \mathbf{0} \end{bmatrix} \begin{bmatrix} \ddot{\mathrm{q}} \\ \boldsymbol{\lambda} \end{bmatrix} = \begin{bmatrix} \mathrm{Q}_e + \mathrm{Q}_v - \mathrm{Kq} \\ \mathrm{Q}_d \end{bmatrix} \tag{4.38}$$

The general mass matrix $\mathbf{M}$ is composed of seven submatrices $\mathbf{M}^i$ corresponding to each link. These submatrices of the rigid and flexible bodies are respectively 3×3 and 5×5, and consequently, the size of general mass matrix is equal to 27×27. The vector of 18 kinematic constraints is formulized in terms of 27 generalized flexible and rigid coordinates defined in Eqs. (4.6) and (4.15). Therefore, the size of Jacobian matrix of the constraints vector is equal to 18×27. Correspondent to each driving constraint, a raw is augmented to the Jacobian matrix in the forward dynamics analysis. Vector $\mathbf{Q}_d$, including the accelerations in terms of velocities, is derived by second derivation respect to the generalized coordinates. The vector $\mathbf{Q}_e + \mathbf{Q}_v - \mathbf{Kq}$ including the generalized forces, centrifugal-Coriolis forces and elastic forces was previously derived. Regarding the motion in the horizontal plan, the gravitational forces are perpendicular to the motion plan and therefore, can be neglected. Moreover, the reaction forces of the joint constraints appear as the generalized forces in terms of the Lagrange multipliers and are not included in the external forces. The only external forces applied to the system are the driving forces of actuators. The first set of actuators applies torque to the revolute joint of the link A_iB_i at point A_i, and the second set applies linear force to the prismatic joint of the flexible link B_iC_i at point B_i.

4.5 Inverse Dynamics Analysis of the Flexible 3RPR Manipulator

4.5.1 Inverse Kinematics and Forward Dynamics of the Rigid Manipulator

The main objective of the inverse dynamics analysis is to find the torques and forces which manipulate the moving platform on a desired path. For this purpose, moving the point p of the end-effector on a desired trajectory is considered as a driving constraint. The path of the moving platform is usually assumed as the imported time-variant numerical parameter. Therefore, the required functions driving the joints have to be defined in terms of time. Associated with each driving constraint, a row is augmented to the constraint Jacobian matrix and a component is added to the Lagrange multipliers. The required applied torques and forces can be determined by solving the equations of motion and finding the Lagrange multipliers. Regarding the main objectives of the analysis which are finding three linear forces of the actuators or three torques of the rotary motors, coordinates of the prismatic joints or angles of the revolute joints have to be found in terms of the moving platform coordinates. First, considering the rigid manipulator, angles $\overline{\theta}_i$ of the revolute joints and coordinates $\overline{\rho}_i$ of the prismatic joints can be found as

$$\begin{aligned} \overline{\theta}_i(t) &= \tan^{-1}\left(\frac{y_P(t) + r\sin(\varphi_i + \phi_P(t)) - Y_{Ai}}{x_P(t) + r\cos(\varphi_i + \phi_P(t)) - X_{Ai}}\right) \\ \overline{\rho}_i(t) &= \frac{x_P(t) + r\cos(\varphi_i + \phi_P(t)) - X_{Ai}}{\cos(\theta_2(t))} - l,\ (i = 1, 2, 3) \end{aligned} \tag{4.39}$$

The desired path is assumed to be a circle whose time-dependent coordinates are written as

$$x_p = -0.05\cos(10\pi t) - 0.06,\quad y_p = 0.05\sin(10\pi t),\quad \phi_p = \pi/4 \tag{4.40}$$

Consequently, driving constraints of the revolute and prismatic joints are respectively defined in Eqs. (4.41) and (4.42) as

$$\mathrm{C}_{19}(\mathbf{q}, t) = \theta_1 - \overline{\theta}_1(t) = 0,\ \mathrm{C}_{20}(\mathbf{q}, t) = \theta_3 - \overline{\theta}_3(t) = 0,\ \mathrm{C}_{21}(\mathbf{q}, t) = \theta_5 - \overline{\theta}_5(t) = 0 \tag{4.41}$$

$$\begin{aligned} \mathrm{C}_{19}(\mathbf{q}, t) &= -(X_{A1} - R_{x2})\cos(\theta_2) - (Y_{A1} - R_{y2})\sin(\theta_2) - \overline{\rho}_1(t) = 0 \\ \mathrm{C}_{20}(\mathbf{q}, t) &= \ \ (X_{A2} - R_{x4})\cos(\theta_4) + (Y_{A2} - R_{y4})\sin(\theta_4) - \overline{\rho}_2(t) = 0 \\ \mathrm{C}_{21}(\mathbf{q}, t) &= -(X_{A3} - R_{x6})\cos(\theta_6) - (Y_{A3} - R_{y6})\sin(\theta_6) - \overline{\rho}_3(t) = 0 \end{aligned} \tag{4.42}$$

All these driving constraints have to be augmented by other 18 constraints which were mentioned previously. Substituting Eq. (4.40) into Eq. (4.39) and then using Eqs. (4.41) and (4.42) yields directly deriving of the constraint equations in terms of the moving platform coordinates. After initializing the conditions of the moving platform consistent with the driving constraints and by numerically solving the system of motion equations, the forward dynamics analysis is formulated. As the result, three forces of the linear actuators or three torques of the motors can be found. However, the process has to be further continued to achieve more reliable results for the flexible manipulator.

4.5.2 Inclusion of the Driving Constraints for Flexible Coordinates

As the result of analysis in previous section, the generalized elastic coordinates of the flexible intermediate links are also obtained. These coordinates are further exploited to form some additional driving constraints. This investigation is conducted to find the precise values of the driving torques or linear forces in path tracking. The flexible manipulator has nine DOFs. Therefore, nine driving constraints have to be defined besides 18 constraints which were mentioned previously. For the first case, the driving constraints vector includes six constraints of the elastic coordinates and three constraints of the angular coordinates written as

$$\begin{array}{ll} C_{19}(\mathbf{q},t) = q_{f1}^{2} - \tilde{q}_{f1}^{2} = 0, & C_{20}(\mathbf{q},t) = q_{f2}^{2} - \tilde{q}_{f2}^{2} = 0, \\ C_{21}(\mathbf{q},t) = q_{f1}^{4} - \tilde{q}_{f1}^{4} = 0, & C_{22}(\mathbf{q},t) = q_{f2}^{4} - \tilde{q}_{f2}^{4} = 0, \\ C_{23}(\mathbf{q},t) = q_{f1}^{6} - \tilde{q}_{f1}^{6} = 0, & C_{24}(\mathbf{q},t) = q_{f2}^{6} - \tilde{q}_{f2}^{6} = 0, \\ C_{25}(\mathbf{q},t) = \theta_1 - \tilde{\theta}_1 = 0, & C_{26}(\mathbf{q},t) = \theta_3 - \tilde{\theta}_3 = 0, \\ \multicolumn{2}{c}{C_{27}(\mathbf{q},t) = \theta_5 - \tilde{\theta}_5 = 0} \end{array} \tag{4.43}$$

For the second case, the driving constraints vector includes six constraints of the elastic coordinates and three constraints of the linear actuators at prismatic joints written as

$$\begin{aligned} &C_{19}(\mathbf{q},t) = q_{f1}^{2} - \tilde{q}_{f1}^{2} = 0, \quad C_{20}(\mathbf{q},t) = q_{f2}^{2} - \tilde{q}_{f2}^{2} = 0, \\ &C_{21}(\mathbf{q},t) = q_{f1}^{4} - \tilde{q}_{f1}^{4} = 0, \quad C_{22}(\mathbf{q},t) = q_{f2}^{4} - \tilde{q}_{f2}^{4} = 0, \\ &C_{23}(\mathbf{q},t) = q_{f1}^{6} - \tilde{q}_{f1}^{6} = 0, \quad C_{24}(\mathbf{q},t) = q_{f2}^{6} - \tilde{q}_{f2}^{6} = 0, \\ &C_{25}(\mathbf{q},t) = -(X_{A1} - R_{x2})\cos(\theta_2) - \left(Y_{A1} - R_{y2}\right)\sin(\theta_2) \\ &\qquad + \left(\left(X_{A1} - \tilde{R}_{x2}\right)\cos\left(\tilde{\theta}_2\right) + \left(Y_{A1} - \tilde{R}_{y2}\right)\sin\left(\tilde{\theta}_2\right)\right) = 0, \\ &C_{26}(\mathbf{q},t) = (X_{A2} - R_{x4})\cos(\theta_4) + \left(Y_{A2} - R_{y4}\right)\sin(\theta_4) \\ &\qquad - \left(\left(X_{A2} - \tilde{R}_{x4}\right)\cos\left(\tilde{\theta}_4\right) + \left(Y_{A2} - \tilde{R}_{y4}\right)\sin\left(\tilde{\theta}_4\right)\right) = 0, \\ &C_{27}(\mathbf{q},t) = -(X_{A3} - R_{x6})\cos(\theta_6) - \left(Y_{A3} - R_{y6}\right)\sin(\theta_6) \\ &\qquad + \left(\left(X_{A3} - \tilde{R}_{x6}\right)\cos\left(\tilde{\theta}_6\right) + \left(Y_{A3} - \tilde{R}_{y6}\right)\sin\left(\tilde{\theta}_6\right)\right) = 0 \end{aligned} \tag{4.44}$$

where the coordinates signed by the symbol $(\tilde{\cdot})$ are time-dependent variables which are obtained from the forward dynamics analysis of the previous section. By imposing the new sets of the driving constraints in the motion equations and solving the corresponding forward dynamics of the flexible manipulator, the Lagrange multipliers associated with the driving constraints are obtained. From this step, the required applied torques or linear forces for tracking the desired trajectory are found. To validate the approach, the obtained actuating forces and torques are applied to the flexible manipulator to check the final trajectory of the end-effector.

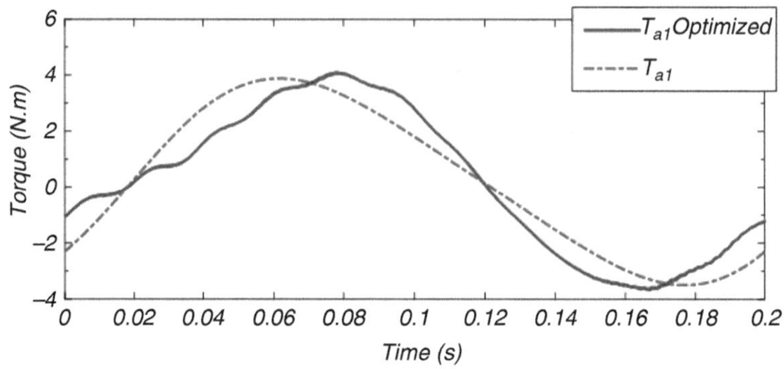

Fig. 4.4 The torque of the rotary motor at joint A_1

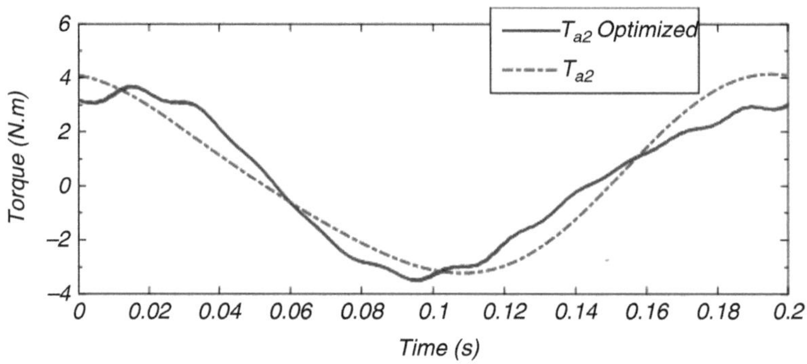

Fig. 4.5 The torque of the rotary motor at jointA_2

For checking the approach, the following structural and geometrical values are assumed for the flexible manipulator

$$
\begin{aligned}
&l_{A_1A_2} = l_{A_2A_3} = l_{A_1A_3} = 750\text{mm}, \mathrm{B}_i\mathrm{C}_i: \quad l_{B_iC_i} = 250\text{mm}, \\
&\quad d_{B_iC_i} = 5\text{mm (diameter)}, \\
&l_{C_1C_2} = l_{C_2C_3} = l_{C_1C_3} = 100\text{mm}, m_7 = 0.2\text{kg} \\
&\rho_{B_iC_i} = 2840\text{kg/m}^3, \quad E_{B_iC_i} = 73\text{ GPa}, \mathrm{A}_i\mathrm{B}_i: \quad d_{\text{out}} = 10\text{mm}, d_{\text{in}} = 8\text{mm}, \\
&\rho_{A_iB_i} = 2840\text{kg/m}^3, \ \overline{\rho}_i = 250\text{mm} \quad \text{(motion limit)}
\end{aligned}
$$

Figures 4.4, 4.5, and 4.6 illustrate the torques of rotary motors resulting from the first step (dashed line) and the second step (bold line). Indeed, the final optimized torques are found from the second step to achieve accurate path tracking. Furthermore, as Figs. 4.7, 4.8 and 4.9 obviously show, the forces of linear actuators resulting from the first and the second steps are respectively illustrated in dashed and bold line styles. Finally, Fig. 4.10 illustrates the final path of the moving platform (bold line) which is compared with the paths resulting from the applied torques and forces of the first step. It can clearly be seen that the final tracked trajectory (bold line) is very close to the desired trajectory defined in Eq. (4.40).

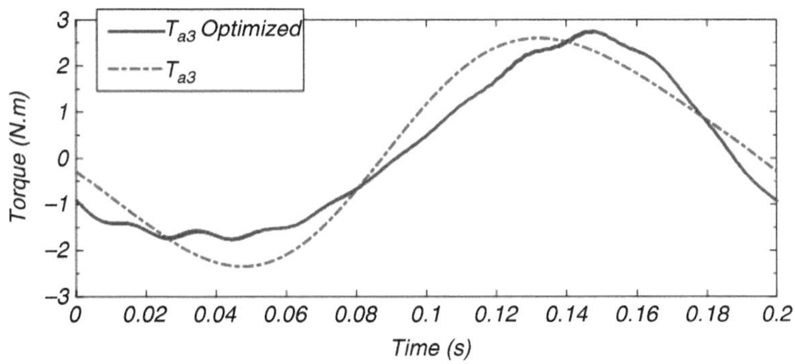

Fig. 4.6 The torque of the rotary motor at jointA_3

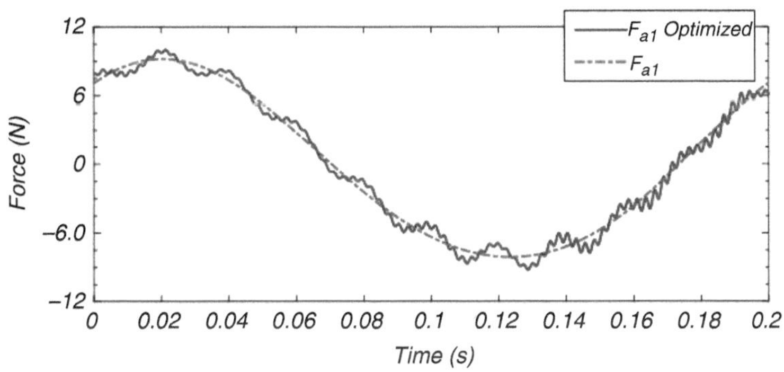

Fig. 4.7 The force of the linear actuator at jointB_1

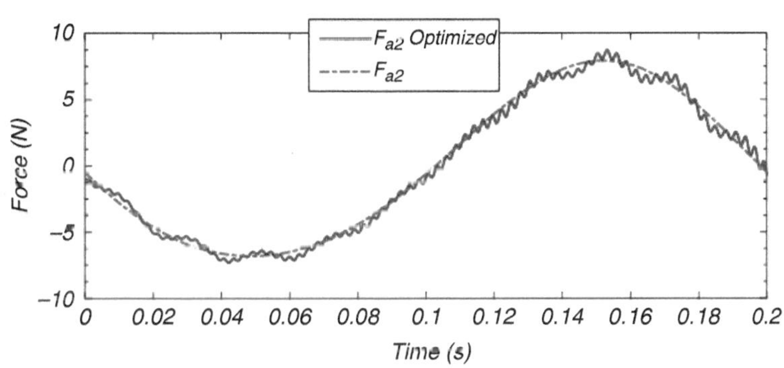

Fig. 4.8 The force of the linear actuator at jointB_2

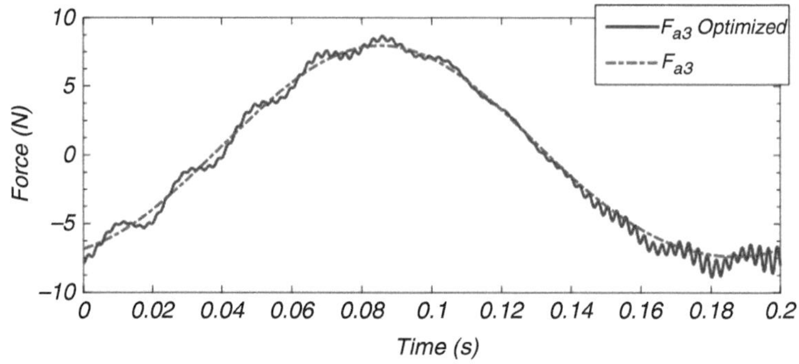

Fig. 4.9 The force of the linear actuator at jointB_3

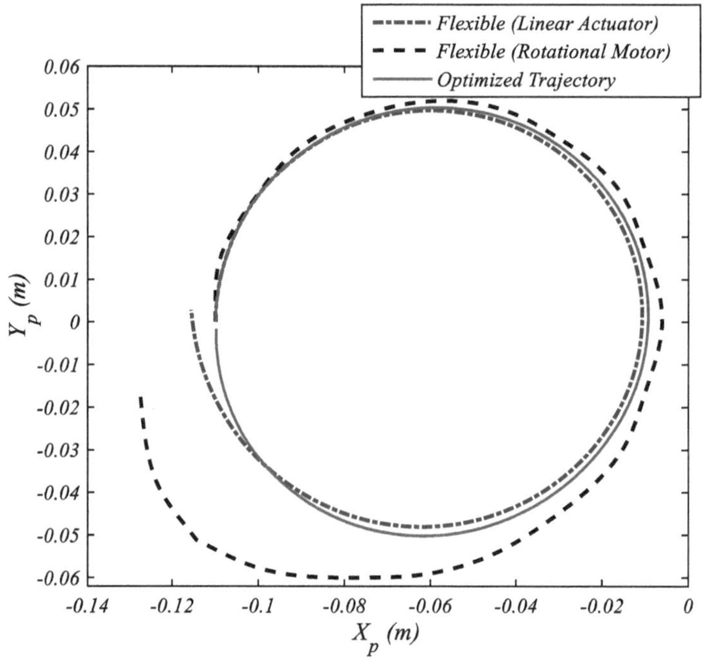

Fig. 4.10 The final path of the moving platform (bold line) compared with the paths resulting from the applied torques and forces of the first step

4.6 Conclusion

A novel procedure for inverse dynamics modeling of flexible multibody systems using a driving constraints approach in the form of forward dynamics problem was presented. The flexible 3RPR parallel manipulator was chosen here as a flexible multibody system with closed kinematic chains to introduce this approach. The equations of motion were derived using the floating frame of reference formulation. Assuming a prescribed trajectory of the end-effector, the generalized coordinates of the rigid manipulator associated with the prismatic and revolute joints were obtained from an inverse kinematics analysis of the manipulator. This solution was further exploited in a forward dynamics analysis of the flexible manipulator to form the required driving constraints for obtaining the approximate values of the actuating forces and torques. Finally, the inverse dynamics analysis of the flexible manipulator was carried out by including some additional driving constraints associated with the generalized elastic coordinates to obtain the high-accuracy approximate values of the actuating forces and torques for tracking the prescribed trajectory by the end-effector. The approach was finally validated by exerting the obtained actuating forces or torques to the simulated flexible manipulator and checking the final trajectory of the end-effector. The results confirmed the accuracy of the proposed approach. To apply this approach in a real application, a feedback controller might also be used to augment the feed forward control calculations presented here.

References

1. Bastos, G., Seifried, R., Brüls, O.: Inverse dynamics of serial and parallel underactuated multibody systems using a DAE optimal control approach. Multibody Syst. Dyn. **30**(3), 359–376 (2013)
2. Lismonde, A., Sonneville, V., Bruls, O.: Inverse dynamics of a flexible 3D robotic arm for a trajectory tracking task. The 4th Joint International Conference on Multibody System Dynamics (IMSD), 29 May-2 June, Montreal, Canada (2016)
3. Rodriguez, G.: Spatial operator approach to flexible manipulator inverse and forward dynamics. International Conference on Robotics and Automation 1990, pp. 845–850 (1990)
4. Ledesma, R., Devasia, S., Bayo, E.: Inverse dynamics of spatial open-chain flexible manipulators with lumped and distributed actuators. J. Field Robot. **11**(4), 327–338 (1994)
5. Mardani, A., Ebrahimi, S.: Computational dynamic modeling and sequential PID controlling of a tendon-based manipulator with highly slender flexible arms. 4th International Conference on Robotics and Mechatronics (ICROM), 26–28 Oct., pp. 542–547, Tehran (2016)
6. Bayo, E., Moulin, H.: An efficient computation of the inverse dynamics of flexible manipulators in the time domain. International Conference on Robotics and Automation 1989, pp. 710–715 (1989)
7. Boyer, F., Khalil, W.: An efficient calculation of flexible manipulator inverse dynamics. Int. J. Robot. Res. **17**(3), 282–293 (1998)
8. Talebi, H.A., Khorasani, K., Patel, R.V.: Neural network based control schemes for flexible link manipulators: simulations and experiments. Neural Netw. **11**(7–8), 1357–1377 (1998)
9. Su, Z., Khorasani, K.: A neural-network-based controller for a single-link flexible manipulator using the inverse dynamics approach. IEEE Trans. Ind. Electron. **48**(6), 1074–1086 (2001)
10. Rahmani, B., Belkheiri, M.: Adaptive neural network output feedback control for flexible multi-link robotic manipulators. Int. J. Control., pp. 1–35 (2018)
11. Sun, Q., Nahon, M., Shart, I.: An inverse dynamics algorithm for multiple flexible-link manipulators. J. Vib. Control. **6**(4), 557–569 (2000)
12. Ledesma, R., Bayo, E.: A Lagrangian approach to the non-causal inverse dynamics of flexible multibody systems: the three-dimensional case. Int. J. Numer. Methods Eng. **37**(19), 3343–3361 (1994)

13. Lismonde, A., Brüls, O., Sonneville, V.: Solving the inverse dynamics of a flexible 3D robot for a trajectory tracking task. International Conference on Methods and Models in Automation and Robotics (MMAR) 2016, pp. 194–199. IEEE (2016)
14. Damaren, C.L.: Approximate inverse dynamics and passive feedback for flexible manipulators with large payloads. IEEE Trans. Robot. Autom. **12**(1), 131–138 (1996)
15. Trautt, T.A., Bayo, E.: Inverse dynamics of flexible manipulators with coulomb friction or backlash and non-zero initial conditions. Dyn. Control. **9**(2), 173–195 (1999)
16. Carrera, E., Serna, M.A.: Inverse dynamics of flexible robots. Math. Comput. Simul. **41**(5–6), 485–508 (1996)
17. Staicu, S.: Inverse dynamics of the 3-PRR planar parallel robot. Robot. Auton. Syst. **57**(5), 556–563 (2009)
18. Staicu, S.: Power requirement comparison in the 3-RPR planar parallel robot dynamics. Mech. Mach. Theory. **44**(5), 1045–1057 (2009)
19. Staicu, S.: Inverse dynamics of the spatial 3-RPS parallel robot. Proc. Rom. Acad. A. **13**(1), 62–70 (2012)
20. Plitea, N., Hesselbach, J., Pisla, D., Raatz, A., Vaida, C., Prodan, B., Dadarlat, R.: Inverse dynamics of a 5-DOF reconfigurable parallel robot. In: 13th World Congress in Mechanism and Machine Science, pp. 19–25. Guanajuato, México (2011)
21. Kordjazi, H., Akbarzadeh, A.: Inverse dynamics of a 3-PRR planar parallel manipulator using natural orthogonal complement. J. Syst. Control Eng. **225**, 258–269 (2011)
22. Asada, H., Ma, Z.D., Tokumaru, H.: Inverse dynamics of flexible robot arms: modeling and computation for trajectory control. J. Dyn. Syst. Meas. Control. **112**(2), 177–185 (1990)
23. Zhaocai, D., Yueqing, Y.: Dynamic modeling and inverse dynamic analysis of flexible parallel robots. Int. J. Adv. Robot. Syst. **5**(1), 13 (2008)
24. Brüls, O., Lismonde, A., Sonneville, V.: Implicit finite element formulation of the inverse dynamics of vibrating robots. Proceedings of the 9th European Nonlinear Dynamics Conference (ENOC), June 25–30, Budapest, Hungary (2017)
25. Moberg, S., Hanssen, S.: Inverse dynamics of flexible manipulators. In: 2009 Conference on Multibody Dynamics 2009, pp. 1–20. Warsaw, Poland (2009)
26. Firoozabadi, A.E., Ebrahimi, S., Amirian, G.: Dynamic characteristics of a 3-RPR planar parallel manipulator with flexible intermediate links. Robotica. **33**(9), 1909–1925 (2015)
27. Ebrahimi, S., Eshaghiyeh-Firoozabadi, A.: Dynamic performance evaluation of serial and parallel RPR manipulators with flexible intermediate links. Iran. J. Sci. Technol., Trans. Mech. Eng. **40**(3), 169–180 (2016)
28. Shabana, A.A.: Dynamics of multibody systems. 4th edn. Cambridge University Press. New York, USA (2013)

Chapter 5
Probabilistic Solutions of the Stretched Beam Systems Formulated by Finite Difference Scheme and Excited by Gaussian White Noise

Guo-Kang Er, Vai Pan Iu, Kun Wang, and Hai-En Du

Abstract The probabilistic solutions of elastic stretched beams are studied when the beam is discretized by finite difference scheme and excited by Gaussian white noise which is fully correlated in space. The nonlinear multi-degree-of-freedom system about the random vibration of stretched beam is formulated by finite difference scheme first. Then the relevant Fokker-Planck-Kolmogorov equation is solved for the probabilistic solutions of the system by the state-space-split and exponential polynomial closure method. Monte Carlo simulation method and equivalent linearization method are also adopted to analyze the probabilistic solutions of the system responses, respectively. Numerical results obtained with the three methods are presented and compared to show the computational efficiency and numerical accuracy of solving the Fokker-Planck-Kolmogorov equation by the state-space-split and exponential polynomial closure method in analyzing the probabilistic solutions of the beams discretized by finite difference scheme and excited by Gaussian white noise. The techniques of using the state-space-split procedure for dimension reduction of the beam systems are discussed through the given beam systems with different space distributions of excitations.

Keywords Stretched beam · Nonlinear random vibration · FPK equation · MDOF system · Finite difference · SSS-EPC method

5.1 Introduction

The vibrations of hinged, axially restrained and elastic stretched beam was studied by many researchers with fruitful results in the past because this beam can find its applications in many areas of science and engineering. However, there are few

G.-K. Er (✉) · V. P. Iu · K. Wang · H.-E. Du
University of Macau, Macau SAR, People's Republic of China
e-mail: gker@umac.mo

E. Zahariev, J. Cuadrado (eds.), *IUTAM Symposium on Intelligent Multibody Systems – Dynamics, Control, Simulation*, IUTAM Bookseries 33,
https://doi.org/10.1007/978-3-030-00527-6_5

studies about the nonlinear random vibrations or probabilistic solutions of the beam when the beam is modeled as a multi-degree-of-freedom (MDOF) nonlinear stochastic dynamical (NSD) system. The hinged, axially restrained and elastic stretched beam was modeled as a MDOF-NSD system with Galerkin's method and then an expression for the joint probability-density function (PDF) of the first N modal amplitudes is obtained under the excitation of uniformly distributed Gaussian white noise which is uncorrelated in space [1]. If the Gaussian white noise is correlated or fully correlated in space as usual in applications, the joint PDF of the first N modal amplitudes is not obtainable because the solution of the relevant Fokker-Planck-Kolmogorov (FPK) equation is not obtainable in this case. The technique of equivalent linearization (EQL) was adopted to study the mean square responses of a simply-supported Bernoulli-Euler beam undergoing moderately large random vibrations [2]. An improved stochastic linearization method was employed to study the modal-amplitude mean-square of stretched beam when the beam is modeled as a MDOF system by Galerkin's method [3]. The random vibrations of the linear beam with flexural boundary conditions were analyzed by dynamic-edge-effect method [4]. The probabilistic solutions of the deflection of stretched beam was studied recently when the beam is modeled as a MDOF system by Galerkin's method and excited by filtered Gaussian white noise [5].

When the system of stretched beam is formulated by finite difference scheme and excited by random noise, it is a MDOF-NSD system. If the excitation is Gaussian white noise or filtered Gaussian white noise, the Fokker-Planck-Kolmogorov (FPK) equation governs the probabilistic solution of the system. However, solving the high-dimensional FPK equation is a challenge because of dimensionality [6], especially for the systems with many degrees of freedom, strongly coupled state variables, strong nonlinearity and many nonlinear terms, such as the nonlinear random vibration of the stretched-beam system formulated by finite difference scheme.

There are three methods popularly adopted for analyzing the MDOF-NSD systems. One of them is the EQL method which was proposed by Booton and further studied and extended by many other researchers thereafter [7–10]. The EQL method works for the weakly nonlinear systems. The Monte Carlo simulation (MCS) is the second method which was proposed by Metropolis and Ulam [11] and further studied by many other researchers thereafter [12, 13]. The challenges inherent with MCS method in analyzing the MDOF-NSD systems with strong nonlinearity are the numerical stability, convergence, round-off error, and requirement for huge sample size. The third method is the cumulant-neglect closure with which the moment equations formulated with FPK equation are solved for the statistical moments when the cumulant-neglect closure is introduced to moment equations to make the hierarchy of moment equations closed to a desired level [14–17]. The exact moments can be obtained with this method for the linear systems excited by additive and multiplicative white noises. Accurate moments can also be obtained for the weakly nonlinear systems excited by additive white noises.

In the last decades, there were some methods proposed for the approximate solutions of high-dimensional FPK equations. The weighted Hermit polynomial functions were adopted to express the approximate PDF for solving the high-dimensional FPK equations [18]. Some 2DOF and 3DOF nonlinear systems were analyzed by this method. The high-dimensional FPK equations were also solved by the method of tensor decomposition [19]. With this method, the FPK equations about the nonlinear systems with one to five degrees of freedom were solved when the Gaussian white noises in the equations of motion are independent each other. Recently, a new method named state-space-split (SSS) method was proposed for the probabilistic solutions of the large MDOF-NSD systems which are solvable by EQL method or for solving the relevant FPK equations in high dimensionality [20, 21]. Some real MDOF systems excited by Gaussian white noise or colored noise have been studied by this method. The idea of the SSS method is that the high-dimensional FPK equation is reduced to the low-dimensional FPK equations which are then solvable with the exponential-polynomial-closure (EPC) method. The whole solution procedure is then called SSS-EPC method. The effectiveness of the SSS-EPC method can be studied for different problems.

In this paper, the SSS-EPC method is further extended and applied to study the probabilistic solutions of stretched beam systems formulated by finite difference scheme. The excitation of Gaussian white noise is applied on the beam laterally and the Gaussian white noise is fully correlated in space. The equation of motion of the stretched beam is a nonlinear partial differential equation in time and space. With the finite difference scheme, the nonlinear partial differential equation governing the motion of the beam is reduced to a MDOF-NSD system. The results obtained by SSS-EPC method are compared with those obtained by EQL method and MCS to examine the computational efficiency and numerical accuracy of SSS-EPC method in analyzing the probabilistic solutions of the stretched-beam systems formulated by finite difference scheme.

5.2 Nonlinear Stochastic Dynamical System of the Stretched Beam

Consider the Euler-Bernoulli beam with pin supports on its two ends and excited by distributed Gaussian white noise as shown in Fig. 5.1.

The governing equation of the beam is

$$\begin{aligned}&\rho A\ddot{Y}(x,t)+c\dot{Y}(x,t)+EIY^{(4)}(x,t)-\tfrac{EA}{2L}Y''(x,t)\\&\textstyle\int_0^L Y'^2(x,t)\,dx=q(x)W(t)\end{aligned} \tag{5.1}$$

Fig. 5.1 The stretched beam

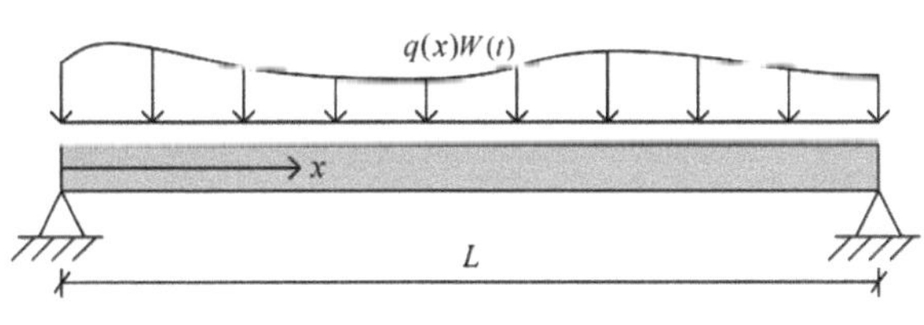

where $Y(x, t)$ is the deflection of the beam at time t at the location with distance x to the left-hand side of the beam; ρ is the mass density of the beam material; c is viscous damping constant; E is the Young's modulus of beam material; I is the moment inertia of beam cross section; A is the area of beam cross section; L is the beam length; $q(x)W(t)$ is the distributed loading laterally applied on the beam, $q(x)$ is a function of x and $W(t)$ is the Gaussian white noise with zero mean and auto-correlation $E[W(t)W(t+\tau)] = S\delta(\tau)$ in which $\delta(\tau)$ is Dirac's delta function and S is a constant representing the power spectral density of $W(t)$.

5.3 Multi-Degree-of-Freedom Nonlinear Stochastic Dynamical System of the Stretched Beam Formulated by Finite Difference Scheme

By central finite difference scheme, Eq. (5.1) can be discretized into the following system as shown in Fig. 5.2.

$$\ddot{Y}_n + \frac{c}{\rho A}\dot{Y}_n + \alpha\left(Y_{n+2} - 4Y_{n+1} + 6Y_n - 4Y_{n-1} + Y_{n-2}\right) - \beta\left(Y_{n+1} - 2Y_n + Y_{n-1}\right)\sum_{i=1}^{N+1}\left(Y_{i+1}^2 + Y_i^2 + Y_{i-1}^2 + Y_{i-2}^2 + Y_{i+1}Y_i - 2Y_{i+1}Y_{i-1} - Y_{i+1}Y_{i-2} - Y_iY_{i-1} - 2Y_iY_{i-2} + Y_{i-1}Y_{i-2}\right) = \frac{q(x_i)}{\rho A}W(t) \qquad (n = 1, 2, \ldots, N) \tag{5.2}$$

where Y_i is the deflection at the point i with distance x_i to the left-hand side of beam; $h = x_{i+1} - x_i$ and $x_i = ih$; $\alpha = \frac{EI}{h^4 \rho A}$; and $\beta = \frac{E}{24Lh^3\rho}$.

For the beam with pin supports on its two ends, the boundary conditions are $Y(0, t) = Y''(0, t) = Y(L, t) = Y''(L, t) = 0$ which can be expressed by finite difference scheme as

$$Y_0 = 0, Y_{-1} = -Y_1, Y_{N+1} = 0, Y_{N+2} = -Y_N \tag{5.3}$$

Introducing the boundary conditions expressed by Eqs. (5.3) to Eq. (5.2), a MDOF-NSD system excited by Gaussian white noise can be formulated. The number of degrees of freedom of the system is N.

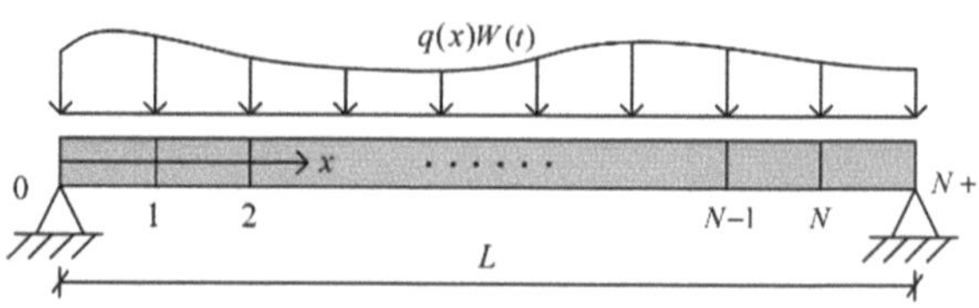

Fig. 5.2 Finite difference model of the stretched beam

5.4 State-Space-Splitting Method and Dimension Reduction for FPK Equation

In order to differentiate the random variables or vectors from the determinate variables or vectors, the random state variables or vectors are denoted by the capital letter and the corresponding deterministic state variables or vectors are denoted by the same letter in lowercase. The summation convention applies in the following discussion unless stated otherwise.

The system governed by Eq. (5.2) under the boundary conditions given by Eq. (5.3) can be generally expressed by the following MDOF-NSD system.

$$\ddot{Y}_i + c_{ij}\dot{Y}_j + h_{i0}\left(\mathbf{Y}\right) = h_i W(t) \qquad i, j = 1, 2, \ldots, n_{\mathbf{y}} \tag{5.4}$$

where $Y_i \in \mathbb{R}$, $(i = 1, 2, \ldots, n_{\mathbf{y}})$, are the components of the vector process $\mathbf{Y} \in \mathbb{R}^{n_{\mathbf{y}}}$; $h_{i0}(\mathbf{Y})$ are the polynomial type of nonlinear functions of $\mathbf{Y}$ and $h_{i0}\left(\mathbf{Y}\right) : \mathbb{R}^{n_{\mathbf{y}}} \to \mathbb{R}$; h_l and c_{ij} are constants; $W(t)$ is the same Gaussian white noise as that in Eq. (5.1).

Setting $Y_i = X_{2i-1}$, $\dot{Y}_i = X_{2i}$, $f_{2i-1} = X_{2i}$, $f_{2i} = -c_{ij}X_{2j} - h_{i0}(\mathbf{X})$, $g_{2i-1} = 0$, $g_{2i} = h_i$ $(i = 1, 2, \ldots, n_{\mathbf{y}})$, and $n_{\mathbf{x}} = 2n_{\mathbf{y}}$, then Eq. (5.4) can be written as follows.

$$\frac{d}{dt}X_i = f_i\left(\mathbf{X}\right) + g_i W(t) \qquad i = 1, 2, \ldots, n_{\mathbf{x}} \tag{5.5}$$

where $\mathbf{X} \in \mathbb{R}^{n_{\mathbf{x}}}$; X_i, $(i = 1, 2, \ldots, n_{\mathbf{x}})$, is the ith component of the state vector process $\mathbf{X}$; $f_i\left(\mathbf{X}\right) : \mathbb{R}^{n_{\mathbf{x}}} \to \mathbb{R}$.

The state vector process $\mathbf{X}$ is Markovian and the PDF $p(\mathbf{x}, t)$ of the Markovian vector is governed by FPK equation. Since the white noise $W(t)$ is Gaussian, the stationary PDF $p(\mathbf{x})$ of the Markovian vector is governed by the following reduced FPK equation [24, 25]:

$$\frac{\partial}{\partial x_j}\left[f_j\left(\mathbf{x}\right) p\left(\mathbf{x}\right)\right] - \frac{1}{2}\frac{\partial^2}{\partial x_i \partial x_j}\left[G_{ij} p\left(\mathbf{x}\right)\right] = 0 \tag{5.6}$$

where $\mathbf{x}$ is the deterministic state vector, $\mathbf{x} \in \mathbb{R}^{n_{\mathbf{x}}}$, and $G_{ij} = g_i g_j S$.

Assume that the following conditions are fulfilled by the solution to Eq. (5.6).

$$\lim_{x_i \to \pm\infty} f_i\left(\mathbf{x}\right) p\left(\mathbf{x}\right) = 0 \quad \text{and} \quad \lim_{x_i \to \pm\infty} \frac{\partial p\left(\mathbf{x}\right)}{\partial x_i} = 0 \qquad i = 1, 2, \ldots, n_{\mathbf{x}} \tag{5.7}$$

It is noted that the deflection and the velocity of the studied beam fulfill the conditions given by Eq. (5.7).

The state vector $\mathbf{X}$ is separated into two parts as $\mathbf{X}_1 \in \mathbb{R}^{n_{\mathbf{x}1}}$ which is referred to as first subspace and $\mathbf{X}_2 \in \mathbb{R}^{n_{\mathbf{x}2}}$ which is referred to as second subspace, i.e., $\mathbf{X} = \{\mathbf{X}_1, \mathbf{X}_2\} \in \mathbb{R}^{n_{\mathbf{x}}} = \mathbb{R}^{n_{\mathbf{x}1}} \times \mathbb{R}^{n_{\mathbf{x}2}}$. The state variables in $\mathbf{X}_1$ are referred to as target state variables. In analyzing the above beam system, define the vector $\mathbf{X}_1$ such that $\mathbf{X}_1 = \{Y_i(t), \dot{Y}_i(t)\}$ or $\mathbf{X}_1 = \{Y_i(t), \dot{Y}_i(t), Y_j(t), \dot{Y}_j(t)\}$ $(j \neq i)$. The equation governing the PDF of $\mathbf{X}_1$ is derived in the following by SSS method. The selection for the components in $\mathbf{X}_1$ will be discussed in the numerical analysis.

Denote the PDF of $\mathbf{X}_1$ as $p_1(\mathbf{x}_1)$. In order to obtain $p_1(\mathbf{x}_1)$, integrating both sides of Eq. (5.6) over $\mathbb{R}^{n_{\mathbf{x}_2}}$ gives

$$\begin{aligned}&\int_{\mathbb{R}^{n_{\mathbf{x}_2}}} \frac{\partial}{\partial x_j}\left[f_j(\mathbf{x})\, p(\mathbf{x})\right] d\mathbf{x}_2 - \frac{1}{2}\int_{\mathbb{R}^{n_{\mathbf{x}_2}}} \frac{\partial^2}{\partial x_i \partial x_j}\left[G_{ij}\, p(\mathbf{x})\right] d\mathbf{x}_2 = 0 \\ &x_i \text{ or } x_j \in \mathbb{R}^{n_{\mathbf{x}}}\end{aligned} \tag{5.8}$$

Under the conditions given by Eq. (5.7), it gives

$$\int_{\mathbb{R}^{n_{\mathbf{x}_2}}} \frac{\partial}{\partial x_j}\left[f_j(\mathbf{x})\, p(\mathbf{x})\right] d\mathbf{x}_2 = 0 \qquad x_j \in \mathbb{R}^{n_{\mathbf{x}_2}} \tag{5.9}$$

and

$$\int_{\mathbb{R}^{n_{\mathbf{x}_2}}} \frac{\partial^2}{\partial x_i \partial x_j}\left[G_{ij}\, p(\mathbf{x})\right] d\mathbf{x}_2 = 0 \qquad x_i \text{ or } x_j \in \mathbb{R}^{n_{\mathbf{x}_2}} \tag{5.10}$$

After integration by part, Eq. (5.8) becomes

$$\begin{aligned}&\int_{\mathbb{R}^{n_{\mathbf{x}_2}}} \frac{\partial}{\partial x_j}\left[f_j(\mathbf{x})\, p(\mathbf{x})\right] d\mathbf{x}_2 - \frac{1}{2}\int_{\mathbb{R}^{n_{\mathbf{x}_2}}} \frac{\partial^2}{\partial x_i \partial x_j}\left[G_{ij}\, p(\mathbf{x})\right] d\mathbf{x}_2 = 0 \\ &x_i, x_j \in \mathbb{R}^{n_{\mathbf{x}_1}}\end{aligned} \tag{5.11}$$

which can be equivalently written as

$$\begin{aligned}&\frac{\partial}{\partial x_j}\left[\int_{\mathbb{R}^{n_{\mathbf{x}_2}}} f_j(\mathbf{x})\, p(\mathbf{x})\, d\mathbf{x}_2\right] - \frac{1}{2}\frac{\partial^2}{\partial x_i \partial x_j}\left[\int_{\mathbb{R}^{n_{\mathbf{x}_2}}} G_{ij}\, p(\mathbf{x})\, d\mathbf{x}_2\right] = 0 \\ &x_i, x_j \in \mathbb{R}^{n_{\mathbf{x}_1}}\end{aligned} \tag{5.12}$$

Collect the terms purely in $\mathbf{x}_1$ in one part and the other terms in the other part. Then $f_i(\mathbf{x})$ is split into two parts as

$$f_j(\mathbf{x}) = f_j^I(\mathbf{x}_1) + f_j^{II}(\mathbf{x}) \tag{5.13}$$

Substituting Eq. (5.13) into Eq. (5.12), it gives

$$\frac{\partial}{\partial x_j}\left[f_j^I(\mathbf{x}_1)\,p_1(\mathbf{x}_1)+\int_{\mathbb{R}^{n_{\mathbf{x}_2}}} f_j^{II}(\mathbf{x})\,p(\mathbf{x})\,d\mathbf{x}_2\right]-\frac{1}{2}\frac{\partial^2[G_{ij}p_1(\mathbf{x}_1)]}{\partial x_i\partial x_j}=0 \qquad (5.14)$$
$$x_i, x_j \in \mathbb{R}^{n_{\mathbf{x}_1}}$$

Express $f_j^{II}(\mathbf{x})$ as $\sum_k f_j^{II}(\mathbf{x}_1,\mathbf{z}_k)$ in which $\mathbf{z}_k \in \mathbb{R}^{n_{\mathbf{z}_k}} \subset \mathbb{R}^{n_{\mathbf{x}_2}}$ and $n_{\mathbf{z}_k}$ is the number of the state variables in $\mathbf{z}_k$. Then Eq. (5.14) becomes

$$\frac{\partial}{\partial x_j}\left[f_j^I(\mathbf{x}_1)\,p_1(\mathbf{x}_1)+\sum_k\int_{\mathbb{R}^{n_{\mathbf{z}_k}}} f_j^{II}(\mathbf{x}_1,\mathbf{z}_k)\,p_k(\mathbf{x}_1,\mathbf{z}_k)\,d\mathbf{z}_k\right]-\frac{1}{2}\frac{\partial^2[G_{ij}p_1(\mathbf{x}_1)]}{\partial x_i\partial x_j}=0$$
$$x_i, x_j \in \mathbb{R}^{n_{\mathbf{x}_1}} \qquad (5.15)$$

in which $p_k(\mathbf{x}_1,\mathbf{z}_k)$ is the joint PDF of $\{\mathbf{X}_1,\mathbf{Z}_k\}$. The summation convention not applies for the indexes k in Eq. (5.15) and in the following discussions.

It is observed from Eq. (5.15) that the coupling of $\mathbf{X}_1$ and $\mathbf{X}_2$ is from the term $f_j^{II}(\mathbf{x}_1,\mathbf{z}_k)\,p_k(\mathbf{x}_1,\mathbf{z}_k)$. Because

$$p_k(\mathbf{x}_1,\mathbf{z}_k)=p_1(\mathbf{x}_1)\,q_k(\mathbf{z}_k|\mathbf{x}_1) \qquad (5.16)$$

where $q_k(\mathbf{z}_k|\,\mathbf{x}_1)$ is the conditional PDF of $\mathbf{Z}_k$ given $\mathbf{X}_1=\mathbf{x}_1$. Substituting Eq. (5.16) into Eq. (5.15), it gives

$$\frac{\partial}{\partial x_j}\left\{\left[f_j^I(\mathbf{x}_1)+\sum_k\int_{\mathbb{R}^{n_{\mathbf{z}_k}}} f_j^{II}(\mathbf{x}_1,\mathbf{z}_k)\,q_k(\mathbf{z}_k|\mathbf{x}_1)\,d\mathbf{z}_k\right]p_1(\mathbf{x}_1)\right\}$$
$$-\frac{1}{2}\frac{\partial^2[G_{ij}p_1(\mathbf{x}_1)]}{\partial x_i\partial x_j}=0\ \ x_i, x_j \in \mathbb{R}^{n_{\mathbf{x}_1}} \qquad (5.17)$$

Unfortunately, $q_k(\mathbf{z}_k|\,\mathbf{x}_1)$ is unknown, so the approximate conditional PDF obtained by EQL is used to replace it. Then Eq. (5.17) becomes the following equation about the approximate PDF of state variables.

$$\frac{\partial}{\partial x_j}\left\{\left[f_j^I(\mathbf{x}_1)+\sum_k\int_{\mathbb{R}^{n_{\mathbf{z}_k}}} f_j^{II}(\mathbf{x}_1,\mathbf{z}_k)\,\overline{q}_k(\mathbf{z}_k|\mathbf{x}_1)\,d\mathbf{z}_k\right]\tilde{p}_1(\mathbf{x}_1)\right\}$$
$$-\frac{1}{2}\frac{\partial^2[G_{ij}\tilde{p}_1(\mathbf{x}_1)]}{\partial x_i\partial x_j}=0\ \ x_i, x_j \in \mathbb{R}^{n_{\mathbf{x}_1}} \qquad (5.18)$$

where $\overline{q}_k(\mathbf{z}_k|\mathbf{x}_1)$ is the approximate conditional PDF of $\mathbf{Z}_k$ obtained by EQL given $\mathbf{X}_1=\mathbf{x}_1$ and $\tilde{p}_1(\mathbf{x}_1)$ is the approximate PDF of $\mathbf{X}_1$. It is noted that the difference between the approximate solution $\tilde{p}_1(\mathbf{x}_1)$ and the exact solution $p_1(\mathbf{x}_1)$ is caused by the approximate conditional PDF $\overline{q}_k(\mathbf{z}_k|\mathbf{x}_1)$. Denote

$$\tilde{f}_j(\mathbf{x}_1) = f_j^I(\mathbf{x}_1) + \sum_k \int_{\mathbb{R}^{n_{\mathbf{z}_k}}} f_j^{II}(\mathbf{x}_1, \mathbf{z}_k)\,\overline{q}_k(\mathbf{z}_k|\mathbf{x}_1)\,d\mathbf{z}_k \tag{5.19}$$

Eq. (5.18) finally becomes

$$\frac{\partial}{\partial x_j}\left[\tilde{f}_j(\mathbf{x}_1)\,\tilde{p}_1(\mathbf{x}_1)\right] - \frac{1}{2}\frac{\partial^2}{\partial x_i \partial x_j}\left[G_{ij}\tilde{p}_1(\mathbf{x}_1)\right] = 0 \qquad x_i, x_j \in \mathbb{R}^{n_{\mathbf{x}_1}} \tag{5.20}$$

Equation (5.20) is the FPK equation governing the approximate joint PDF of the target state variables in subspace $\mathbb{R}^{n_{\mathbf{x}_1}}$. It is seen that $\mathbf{X}_1$ only contains two state variables Y_i and $\dot{Y}_i$ at the node i or four state variables Y_i and $\dot{Y}_i$ at the node i and Y_j and $\dot{Y}_j$ at the node j ($j \neq i$). Therefore, the dimension-reduced FPK equation is two-dimensional or four-dimensional. In the following numerical analysis, the EPC method is used to solve Eq. (5.20) [22, 23].

5.5 Numerical Analysis

Example 1 Consider the stretched beam with pin supports on its two ends. The uniformly distributed load is applied over the whole length of the beam as shown by Fig. 5.3. The beam parameters are given by $\boldsymbol{L} = \mathbf{5}\boldsymbol{m}$, $\boldsymbol{E} = \mathbf{2.1} \times \mathbf{10^{11}}\boldsymbol{pa}$, $\boldsymbol{I} = \mathbf{2.17} \times \mathbf{10^{-4}}\boldsymbol{m^4}$, $\boldsymbol{A} = \mathbf{8.6} \times \mathbf{10^{-3}}\boldsymbol{m^2}$, $\boldsymbol{\rho} = \mathbf{7.85} \times \mathbf{10^3}\boldsymbol{kg/m^3}$, $\boldsymbol{c} = \mathbf{10^3}$, and $\boldsymbol{qW(t)} = \mathbf{2.5} \times \mathbf{10^4}\boldsymbol{W(t)N/m}$ with $\boldsymbol{S} = \mathbf{1}$. The number of unknowns N of the system formulated by finite difference scheme is 9. The equations of motion of the system formulated by finite difference scheme are

$$\begin{aligned}
&\ddot{Y}_n + \tfrac{c}{\rho A}\dot{Y}_n + \alpha\left(Y_{n+2} - 4Y_{n+1} + 6Y_n - 4Y_{n-1} + Y_{n-2}\right) - \beta\left(Y_{n+1} - 2Y_n + \right. \\
&\left. Y_{n-1}\right)\textstyle\sum_{i=1}^{N+1}\left(Y_{i+1}^2 + Y_i^2 + Y_{i-1}^2 + Y_{i-2}^2 + Y_{i+1}Y_i - 2Y_{i+1}Y_{i-1}\right. \\
&- Y_{i+1}Y_{i-2} - Y_iY_{i-1} - \\
&\left. 2Y_iY_{i-2} + Y_{i-1}Y_{i-2}\right) = \tfrac{q}{\rho A}W(t) \qquad (n = 1, 2, \ldots, 9)
\end{aligned} \tag{5.21}$$

with the boundary conditions $Y_0 = 0$, $Y_{-1} = -Y_1$, $Y_{10} = 0$, $Y_{11} = -Y_9$.

The formulated system is a 9-DOF system excited by fully correlated Gaussian white noises. From the dimension-reduction procedure of SSS method, it is known

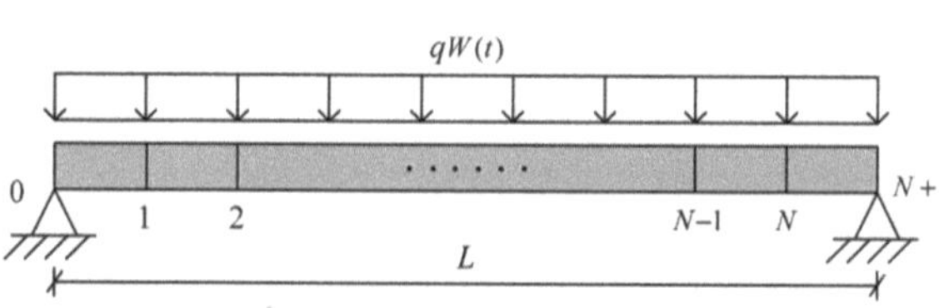

Fig. 5.3 Finite difference model of beam under uniformly distributed load

that the subspace of state variables must be determined first. In other words, the state variables of which the PDF solution is governed by the approximate FPK equation must be determined first. There are two ways for selecting the required state variables for the beam with uniformly distributed load. One is that the displacement Y_i and its corresponding velocity $\dot{Y}_i$ are selected as the target state variables in the first subspace. The dimension-reduced FPK equation to be solved by EPC method is two-dimensional in this case. If the target state variables are selected in this way, numerical analysis shows that the PDFs and the logarithm of PDFs of the Y_i $(i = 2, 3, \ldots, 8)$ are all very close to MCS, but those of Y_1 and Y_9 have some deviation from MCS though they are still much better than the results from EQL. It is observed that $\dot{Y}_i$ $(i = 1, 2, \ldots, 9)$ are all Gaussian and the PDFs of them obtained by SSS-EPC, EQL and MCS are the same. Another way for selecting the target state variables in the first subspace is that the state variables Y_i and $\dot{Y}_i$ in the ith degree of freedom and the state variables Y_j and $\dot{Y}_j$ in the jth $(j \neq i)$ degree of freedom are selected to formulate a four-dimensional approximate FPK equation with SSS method in order to obtain the PDF of Y_i and $\dot{Y}_i$. In order to improve the PDF solution of Y_1, the state variables Y_1 and $\dot{Y}_1$ can be combined with the state variables at another point, e.g., Y_5 and $\dot{Y}_5$ to formulate the target state variables in the first subspace. Then the obtained results about Y_1 can be much improved and very close to MCS as shown in Fig. 5.4. In Fig. 5.4 and the following figures, m_{y_i} and σ_{Y_i} denote the mean value and standard deviation of Y_i obtained by EQL, respectively. For the given examples, $m_{y_i} = 0$.

The PDFs and logarithm of PDFs of the deflection Y_5 in the middle of the beam obtained by SSS-EPC, MCS, and EQL, respectively, are shown and compared in Fig. 5.5a, b, respectively. $m_{y_5} = 0$ and $\sigma_{y_5} = 0.181$ m in this example. The number of samples used in MCS is 2×10^8. The PDFs and the logarithm of PDFs of the deflections Y_i $(i = 2, 3, 4)$ obtained by various methods are also shown and compared in Figs. 5.6, 5.7, and 5.8, respectively. It is seen from these figures that the results obtained by SSS-EPC are close to MCS while those obtained by EQL deviate a lot

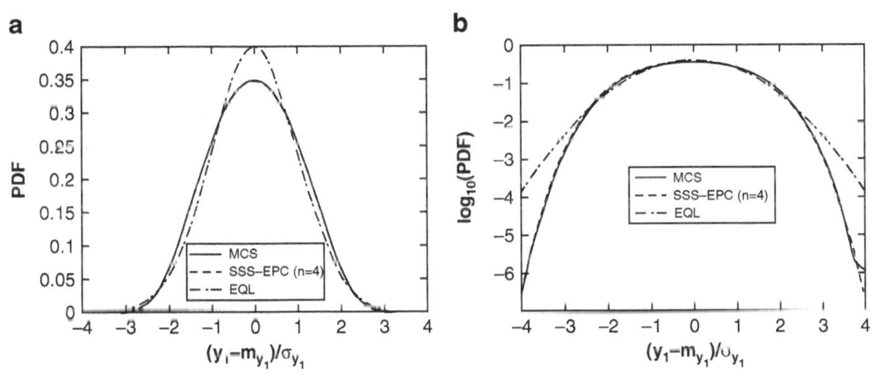

Fig. 5.4 (**a**) The PDFs of the deflection Y_1 under uniformly distributed load over the whole beam; (**b**) The logarithm of PDFs of the deflection Y_1 under uniformly distributed load over the whole beam

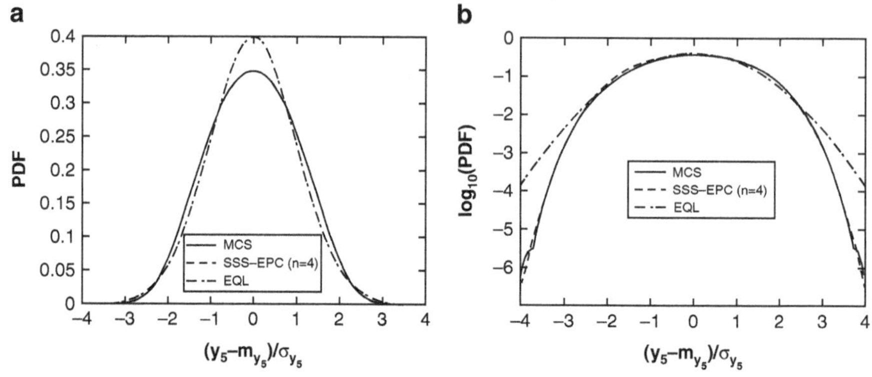

Fig. 5.5 (**a**) The PDFs of the deflection Y_5 under uniformly distributed load over the whole beam; (**b**) The logarithm of PDFs of the deflection Y_5 under uniformly distributed load over the whole beam

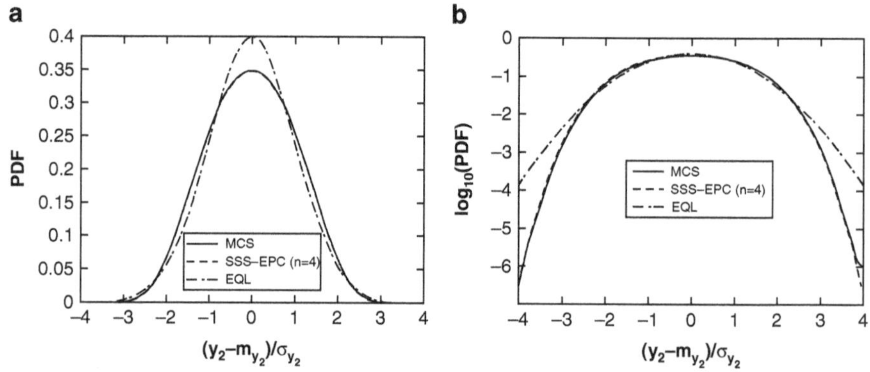

Fig. 5.6 (**a**) The PDFs of the deflection Y_2 under uniformly distributed load over the whole beam; (**b**) The logarithm of PDFs of the deflection Y_2 under uniformly distributed load over the whole beam

from MCS. Under the same computer running environment, the computational time needed by MCS is about 400 times of that needed by SSS-EPC method for this 9-DOF system. The computational time needed by SSS-EPC method is mainly spent on the EQL procedure. The relative computational time between MCS and SSS-EPC can increase exponentially as the number of degrees of system freedom increases.

Numerical results show that $\dot{Y}_i$ $(i = 1, 2, \ldots, 5)$ are all Gaussian and the PDFs of them obtained by SSS-EPC, EQL and MCS are also the same, they are not presented here. Because of the system symmetry, the results of Y_{10-i} and $\dot{Y}_{10-i}$ $(i = 1, 2, 3, 4)$ are the same as those of Y_i and $\dot{Y}_i$ $(i = 1, 2, 3, 4)$, respectively.

Example 2 Consider the stretched beam with pin supports on its two ends. The distributed load is only applied between point 4 and point 5 of the beam as shown by

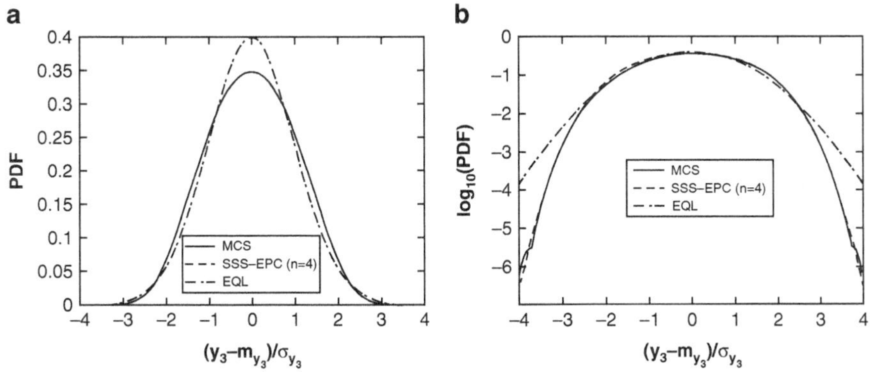

Fig. 5.7 (**a**) The PDFs of the deflection Y_3 under uniformly distributed load over the whole beam; (**b**) The logarithm of PDFs of the deflection Y_3 under uniformly distributed load over the whole beam

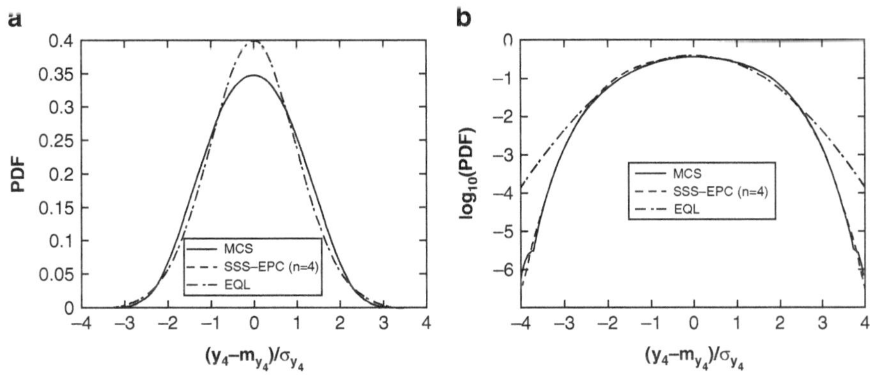

Fig. 5.8 (**a**) The PDFs of the deflection Y_4 under uniformly distributed load over the whole beam; (**b**) The logarithm of PDFs of the deflection Y_4 under uniformly distributed load over the whole beam

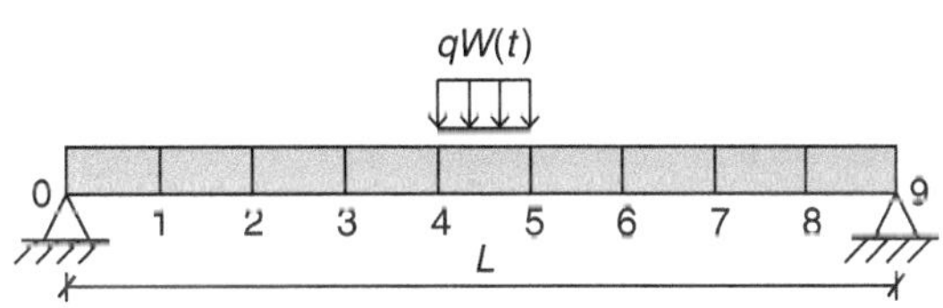

Fig. 5.9 Finite difference model of beam under partially distributed load in the middle of the beam

Fig. 5.9, which can be considered as a point load applied in the middle of beam. The beam parameter values are all be same as those in Example 1 except $\boldsymbol{q} = \mathbf{1.352} \times \mathbf{10^5}$ at points 4 and 5 but $\boldsymbol{q} = \mathbf{0}$ elsewhere. The number of unknowns $\boldsymbol{N}$ of the system formulated by finite difference scheme is 8.

The equations of motion of the system formulated by finite difference scheme are

$$\ddot{Y}_n+\frac{c}{\rho A}\dot{Y}_n+\alpha\left(Y_{n+2}-4Y_{n+1}+6Y_n-4Y_{n-1}+Y_{n-2}\right)-\beta\left(Y_{n+1}-2Y_n+Y_{n-1}\right)$$
$$\sum_{i=1}^{N+1}\left(Y_{i+1}^2+Y_i^2+Y_{i-1}^2+Y_{i-2}^2+Y_{i+1}Y_i-2Y_{i+1}Y_{i-1}-Y_{i+1}Y_{i-2}\right.$$
$$\left.-Y_iY_{i-1}-2Y_iY_{i-2}+Y_{i-1}Y_{i-2}\right)=\frac{q}{2\rho A}W(t) \qquad (n=4,5) \tag{5.22}$$

and

$$\ddot{Y}_n+\frac{c}{\rho A}\dot{Y}_n+\alpha\left(Y_{n+2}-4Y_{n+1}+6Y_n-4Y_{n-1}+Y_{n-2}\right)$$
$$-\beta\left(Y_{n+1}-2Y_n+Y_{n-1}\right)\sum_{i=1}^{N+1}\left(Y_{i+1}^2+Y_i^2+Y_{i-1}^2+Y_{i-2}^2+Y_{i+1}Y_i-2Y_{i+1}Y_{i-1}\right.$$
$$\left.-Y_{i+1}Y_{i-2}-Y_iY_{i-1}-2Y_iY_{i-2}+Y_{i-1}Y_{i-2}\right)=0 \qquad (n=1,2,3,6,7,8) \tag{5.23}$$

with the boundary conditions $Y_0=0$, $Y_{-1}=-Y_1$, $Y_9=0$, $Y_{10}=-Y_8$.

The formulated system is an 8-DOF system. There are Gaussian white noises only in the fourth and fifth equations of motion and the Gaussian white noises in the fourth and fifth equations are fully correlated.

In this case, it is not enough to select the Y_i and $\dot{Y}_i$ $(i \neq 4$ and $5)$ only as the target state variables in the first subspace for the dimension reduction with SSS procedure. If only the state variables Y_i and $\dot{Y}_i$ $(i \neq 4$ and $5)$ are selected as the target state variables in the first subspace, there is no diffusion term in the dimension-reduced FPK equation. Consequently, the formulated two-dimensional approximate FPK equation is not solvable. In this case, the state variables Y_i and $\dot{Y}_i$ $(i \neq 4$ and $5)$ and the state variables Y_j and $\dot{Y}_j$ $(j = 4$ or $5)$ are selected as the target state variables to formulate a four-dimensional first subspace for dimension reduction in the SSS procedure.

Selectingthe state variables Y_i and $\dot{Y}_i$ $(i = 1, 2,$ or $3)$ and Y_j and $\dot{Y}_j$ $(j = 4)$ as target state variables, the results obtained by various methods are shown and compared in Figs. 5.10, 5.11, 5.12, and 5.13, respectively. It is seen from these figures that the results obtained by SSS-EPC method are close to MCS but the results from EQL deviate from MCS a lot. The sample size in MCS is 2×10^8 and the computational time needed by MCS is about 320 times of that needed by SSS-EPC method. It is observed that the results from MCS have some fluctuation in the tails of the PDF of Y_i because of the limited number of samples. More samples and then more computation effort are needed to smoothen the tails of the PDF

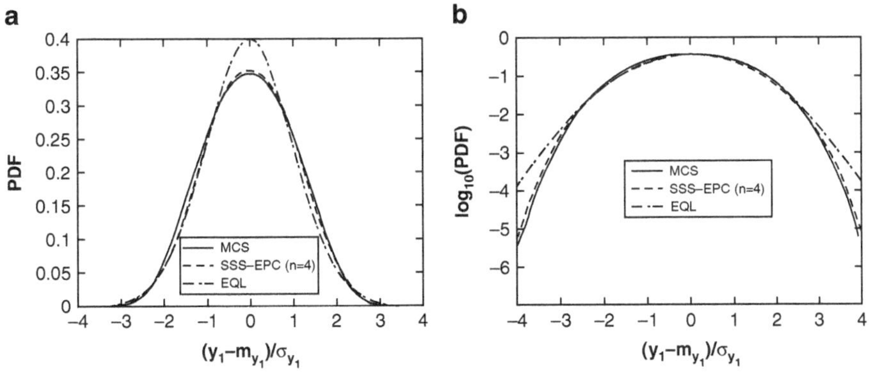

Fig. 5.10 (**a**) The PDFs of the deflection Y_1 under partially distributed load in the middle of the beam; (**b**) The logarithm of PDFs of the deflection Y_1 under partially distributed load in the middle of the beam

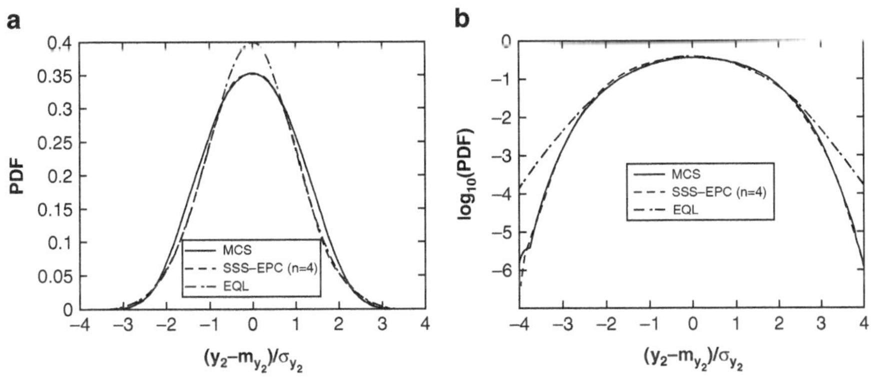

Fig. 5.11 (**a**) The PDFs of the deflection Y_2 under partially distributed load in the middle of the beam; (**b**) The logarithm of PDFs of the deflection Y_2 under partially distributed load in the middle of the beam

solutions obtained by MCS. The ratio of the computational time needed by MCS and that needed by the SSS-EPC method may increase exponentially as the number of equations of motion increases. Numerical results show that $\dot{Y}_i$ is Gaussian and the PDFs of $\dot{Y}_i$ obtained by SSS-EPC, EQL and MCS are all the same. Therefore, they are not presented here. Because of the system symmetry, the PDFs of Y_{9-i} and $\dot{Y}_{9-i}$ are the same as those of Y_i and $\dot{Y}_i$ $(i = 1, 2, 3, 4)$, respectively.

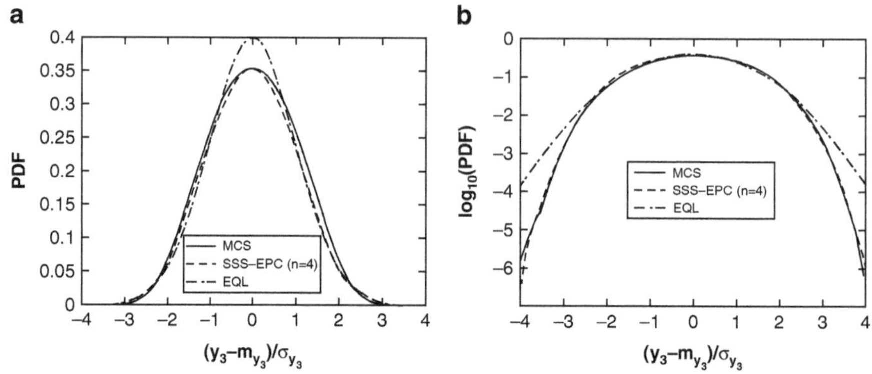

Fig. 5.12 (**a**) The PDFs of the deflection Y_3 under partially distributed load in the middle of the beam; (**b**) The logarithm of PDFs of the deflection Y_3 under partially distributed load in the middle of the beam

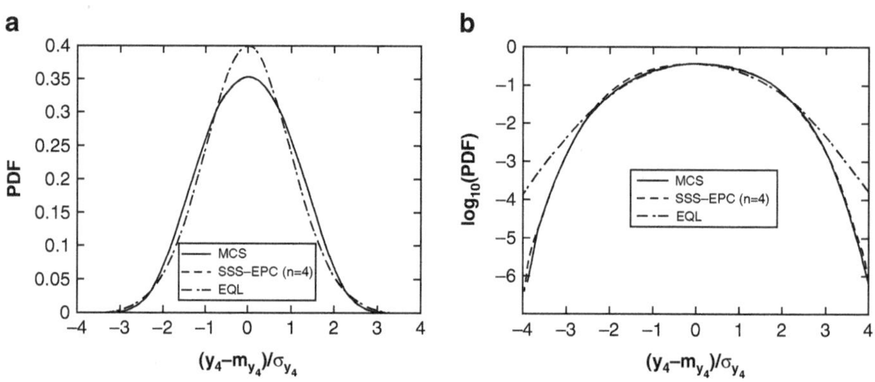

Fig. 5.13 (**a**) The PDFs of the deflection Y_4 under partially distributed load in the middle of the beam; (**b**) The logarithm of PDFs of the deflection Y_4 under partially distributed load in the middle of the beam

5.6 Conclusions

The equations of motion of the stretched beam with pin supports at its two ends are formulated by finite difference scheme when the beam is excited by the distributed Gaussian white noise which is fully correlated in space. The SSS-EPC method is extended to analyze the probabilistic solutions of the beam discretized by finite difference scheme. Both the beams excited by the uniformly distributed load over the whole length of the beam and the locally distributed load applied in the middle of the beam are analyzed. When the beam is modeled as an N-DOF system, the dimension-reduction procedure of the SSS method is adopted to make the 2N-dimensional FPK equation governing the PDF solutions of the beam reduced to some two-dimensional or four-dimensional FPK equations. Then the EPC method is adopted to solve the low-dimensional FPK equations. The techniques about selecting the target state variables in the SSS procedure are presented and discussed for the discretized beam systems. The first subspace in the SSS dimension-reduction procedure can be two-dimensional in some cases but it must be four-dimensional in the others for the stretched beam. The effectiveness and efficiency of the SSS-EPC method are examined by comparing the results obtained by MCS and EQL method. From numerical analysis it is observed that the SSS-EPC method works well for obtaining the PDFs of the beam deflections at all points. The computational time needed by MCS for the analyzed 8 or 9-DOF beam systems is about hundreds times that needed by SSS-EPC method when the sample size is 2×10^8. The solution obtained from EQL is far from being acceptable for the nonlinear random vibration problem of stretched beam.

Acknowledgement The results presented in this paper were obtained under the supports of the Science and Technology Development Fund of Macau (Grant No. 042/2017/A1) and the Research Committee of University of Macau (Grant No. MYRG2018-00116-FST).

References

1. Herbert, R.E.: Random vibrations of a nonlinear elastic beam. J. Acoust. Soc. Am. **36**, 2090–2094 (1964)
2. Herbert, R.E.: On the stresses in a nonlinear beam subject to random excitation. Int. J. Solids Struct. **1**, 235–242 (1965)
3. Fang, J., Elishakoff, I., Caimi, R.: Nonlinear response of a beam under stationary random excitation by improved stochastic linearization method. Appl. Math. Mod. **9**, 106–111 (1995)
4. Elishakoff, I., Lin, Y.K., Zhu, L.P.: Random vibration of uniform beams with varying boundary conditions by the dynamic-edge-effect method. Comput. Methods Appl. Mech. Eng. **121**, 59–76 (1995)
5. Er, G.K.: The probabilistic solutions of some nonlinear stretched beams excited by filtered white noise. In: Proppe, C. (ed) IUTAM Symposium on Multiscale Problems in Stochastic Mechanics 2012. Procedia IUTAM, Vol. **6**, pp. 141–150. Elsevier (2013)
6. Masud, A., Bergman, L.A.: Solution of the four dimensional Fokker-Planck equation: still a challenge. In: Augusti, G., Schuëller, G.I., Ciampoli, M. (eds.) Proceedings of ICOSSAR'2005, pp. 1911–1916, Millpress Science Publishers, Rotterdam (2005)
7. Booton, R.C.: Nonlinear control systems with random inputs. IRE Trans Circ Theory. **CT-1**(1), 9–18 (1954)
8. Kazakov, I.E.: Generaliztion of the method of stochastical linearization to multi-dimensional systems. Autom Remote Control. **26**, 1202–1206 (1965)
9. Socha, L., Soong, T.T.: Linearization in analysis of nonlinear stochastic systems. ASME J. Appl. Mech. Rev. **44**, 399–422 (1991)
10. Proppe, C., Pradlwater, H.J., Schuëller, G.I.: Equivalent linearization and Monte Carlo simulation in stochastic dynamics. Prob. Eng. Mech. **18**, 1–15 (2003)
11. Metropolis, N., Ulam, S.: Monte Carlo method. J. Am Stat Assoc. **14**, 335–341 (1949)
12. Harris, C.J.: Simulation of multivariable nonlinear stochastic systems. Int. J. Num. Meth. Eng. **14**, 37–50 (1979)
13. Kloeden, P.E., Platen, E.: Numerical Solution of Stochastic Differential Equations. Springer, Berlin (1992)

14. Wu, W.F., Lin, Y.K.: Cumulant-neglect closure for nonlinear oscillators under random parametric and external excitations. Int. J. Non-Linear Mech. **19**, 349–362 (1984)
15. Ibrahim, R.A., Soundararajan, A., Heo, H.: Stochastic response of nonlinear dynamic systems based on a non-Gaussian closure. ASME J. Appl. Mech. **52**, 965–970 (1985)
16. Sun, J.-Q., Hsu, C.S.: Cumulant-neglect closure method for nonlinear systems under random excitations. ASME J. Appl. Mech. **54**, 649–655 (1987)
17. Hasofer, A.M., Grigoriu, M.: A new perspective on the moment closure method. ASME J. Appl. Mech. **62**, 527–532 (1995)
18. Martens, W., von Wagner, U., Mehrmann, V.: Calculation of high-dimensional probability density functions of stochastically excited nonlinear mechanical systems. Nonlinear Dyn. **67**, 2089–2099 (2012)
19. Sun, Y., Kumar, M.: Numerical solution of high dimensional stationary Fokker-Planck equations via tensor decomposition and Chebyshev spectral differentiation. Comput Math Appl. **67**, 1961–1977 (2014)
20. Er, G.K.: Methodology for the solutions of some reduced Fokker-Planck equations in high dimensions. Ann. Phys. (Berlin). **523**(3), 247–258 (2011)
21. Er, G.K., Iu, V.P.: A new method for the probabilistic solutions of large-scale nonlinear stochastic dynamic systems. In: Zhu, W.Q., Lin, Y.K., Cai, G.Q. (eds.) Nonlinear Stochastic Dynamics and Control, IUTAM Book Series, Vol. **29**, pp. 25–34. Springer, New York (2011)
22. Er, G.K.: An improved closure method for analysis of nonlinear stochastic systems. Nonlinear Dyn. **17**(3), 285–297 (1998)
23. Er, G.K.: The probabilistic solutions to non-linear random vibrations of multi-degree-of-freedom systems. ASME J. Appl. Mech. **67**(2), 355–359 (2000)
24. Soong, T.T.: Random Differential Equations in Science and Engineering. Academic, New York (1973)
25. Lin, Y.K., Cai, G.Q.: Probabilistic Structural Dynamics. McGraw-Hill, New York (1995)

Chapter 6
O(n) Algorithm for Elastic Link/Joint Robots with End-Effector Contact

Hubert Gattringer, Andreas Müller, Florian Pucher, and Alexander Reiter

Abstract This paper deals with the dynamical modeling of flexible multibody systems like elastic robots that go in contact with the environment. Models for elastic systems have a large degree of freedom leading to longer calculation times for solving the equations of motion (EOM). Conventionally, this includes the inversion of the mass matrix with a cubic run time complexity $O(n^3)$. By using a subsystem formulation and the Projection Equation an $O(n)$ algorithm can be formulated that significantly reduces the simulation time. Additional contacts with the environment can be included in the equations of motion by the corresponding constraint Jacobian and the contact forces. For the explicit calculation of these forces, normally the inverse of the mass matrix is needed again. A novel algorithm to avoid this inversion is presented. Therein, the contact forces are calculated by additional runs of the same $O(n)$ algorithm that is used without contact. The transition phase between different contact states is treated with the help of Newton's impact law, again avoiding the inversion of the mass matrix. Simulation results for an elastic robot show the effectiveness of the proposed algorithms.

Keywords Dynamical modeling · Elastic multibody systems · Ritz approximation · O(n) algorithm · Contact

6.1 Introduction

Flexible multibody systems (FMBS), such as the flexible link robot in Fig. 6.1 are becoming increasingly relevant for industrial applications since they allow for reduced energy consumption and higher acceleration performance as well as lightweight design and thus lower manufacturing costs at the same time. However, the simulation and control of such robots is more complicated [6, 19] than that

---

H. Gattringer (✉) · A. Müller · F. Pucher · A. Reiter
Institute of Robotics, Johannes Kepler University Linz, Linz, Austria

E. Zahariev, J. Cuadrado (eds.), *IUTAM Symposium on Intelligent Multibody Systems – Dynamics, Control, Simulation*, IUTAM Bookseries 33,
https://doi.org/10.1007/978-3-030-00527-6_6

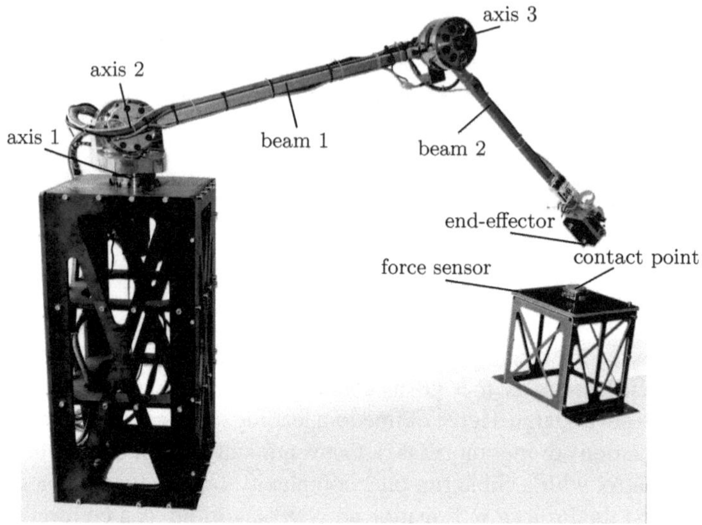

Fig. 6.1 Photo of the robotic system under consideration

of rigid robots since they typically have more degrees of freedom (DoF) due to the elasticity modeling. Therefore efficient algorithms for simulation are necessary. $O(n)$ algorithms for the efficient solution of the equations of motion (EoM) for the generalized accelerations have a long tradition, see e.g. [3, 16, 21]. Recent works are [1, 15, 17] and [8]. A well structured procedure can be introduced using the Projection Equation in subsystem representation, see [5, 9, 14]. Recently, there appear a lot of robotic applications where the robot is in contact with the environment e.g. for grinding or deburring tasks. Typically, industrial robots are used for such tasks leading to high contact forces due to the stiff contact. The usage of elastic robots can drastically reduce these forces. For any type of robot, the EoM can be enhanced by the corresponding constraint Jacobian multiplied by the contact forces. For the calculation of the contact forces, the inverse of the mass matrix is needed resulting in high computation times for many DoF. Again recursive formulations are preferable. In [4] and [11] approaches for the simulation of such closed kinematic systems are discussed. In the present paper a novel approach for the calculation of the contact forces is introduced. The key idea is to separately treat each component of the contact constraints and to calculate the corresponding part of the contact forces using parts of an $O(n)$ algorithm. More precisely, the constraint forces corresponding to the individual columns of the constraint Jacobian are determined. During the transition between the no-contact and contact the velocity of the system jumps [13, 18]. The proposed method is applicable for tree structured multibody systems and parallel kinematic systems, where the kinematic loops are closed by the corresponding contact forces. The momentum balance calculated over the impact interval leads to a system of equations for the velocity jump and the impulsive contact forces. Direct solution of this system involves inversion of the mass matrix. In this work again an algorithm is presented that avoids the inversion of the mass matrix. Simulation results are presented in this paper for which the experimental setup is shown in Fig. 6.1. It consists of three driven axis each of them is equipped with a synchronous motor and a Harmonic Drive gear. The elastic beams are made of steel and have a square hollow cross profile at a length of about 1 m. The maximum elastic deformations at the end-effector are in the range of 5 cm.

The paper is organized as follows. Sections 6.2 and 6.3 show the kinematical and dynamical model for flexible multibody systems with tree structure, respectively. In this chapter also an $O(n)$ algorithm for the calculation of minimal accelerations is presented. In this context, mechanical subsystems consisting of motors, gears, elastic links and tip masses are introduced. In Sect. 6.4 these FMBS models are enhanced to include contact at the end-effector. Also for this case, an $O(n)$ algorithm is worked out. Additionally, the transition phase between free system and system in contact is treated. Simulation results for an elastic robot that enters stiff (steel-steel) contact are discussed in Sect. 6.5.

6.2 Kinematic Modeling

6.2.1 Bernouilli Euler Beam

Flexible joint/link robots can be separated into subsystems, each consisting of geared drives, flexible links and tip bodies, see Fig. 6.2. While details about the subsystem modeling of flexible link/joint robots can be found in [5] and [9], this chapter summarizes the relevant steps in order to make the reader familiar with the notation.

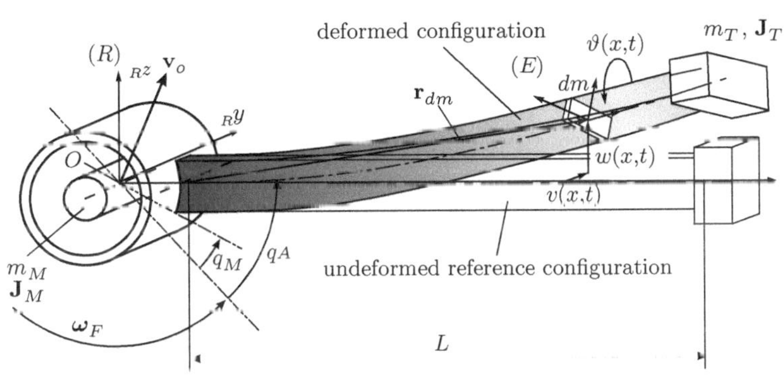

Fig. 6.2 Subsystem motor, elastic gear, elastic arm, tip body

As shown in Fig. 6.2 the motor with mass m_M and inertia tensor $\mathbf{J}_M$ is rigidly connected to the previous subsystem at point O. The subsystem is guided by translational velocities $\mathbf{v}_o$ and rotational velocities $\boldsymbol{\omega}_F$. The relative motor rotation is measured by the motor angle q_M. Between motor and link there is an elastic gear that is modeled by a linear spring with stiffness c_G. The gear output angle is denoted with q_A. The tip body is parameterized by the mass m_T and the inertia tensor $\mathbf{J}_T$. The beam is considered as a one-dimensional continuum. Denote with $\mathbf{r}_{dm}$ the position vector and with $\boldsymbol{\varphi}_{dm}$ the vector of orientation parameters of a cross section element (see Fig. 6.2) w.r.t. the undeformed reference configuration

$$ {}_R\mathbf{r}_{dm} = \begin{pmatrix} x \\ v(x,t) \\ w(x,t) \end{pmatrix}, \quad {}_R\boldsymbol{\varphi}_{dm} = \begin{pmatrix} \vartheta(x,t) \\ -w'(x,t) \\ v'(x,t) \end{pmatrix}. \tag{6.1} $$

The coordinate along the beam axis is x, and the beam has the length L.

The deflections in local y and z directions are denoted v, w, respectively, while ϑ is the torsional angle. All values are described in coordinate frame R moving with the undeformed beam. A prime indicate spatial derivatives w.r.t x: $(.)' = \frac{\partial}{\partial x}$. The rotation matrix $\mathbf{R}_{ER}$ between the reference frame R and an element fixed frame E, cf. Fig. 6.2, is approximated assuming small angles $\boldsymbol{\varphi}_{dm}$ by

$$ \mathbf{R}_{ER} = \mathbf{I} - \tilde{\boldsymbol{\varphi}}_{dm}, \tag{6.2} $$

wherein the tilde operator $\tilde{()}$ corresponds to the vector cross product $\mathbf{a}\times\mathbf{b} = \tilde{\mathbf{a}}\mathbf{b}$ and $\mathbf{I}$ is the identity matrix. The variables v, w, ϑ depend on position x and time t leading to partial differential equations of motion. For the later on dynamical simulation, an approximation method has to be applied. In this paper, the Ritz approximation method is used, where the distributed parameters are separated into chosen spatial shape functions $\mathbf{v}$, $\mathbf{w}$, $\boldsymbol{\vartheta}$ and time dependent Ritz functions $\mathbf{q}$

$$ v(x,t) = \mathbf{v}(x)^T \mathbf{q}_v(t) \tag{6.3} $$

$$ w(x,t) = \mathbf{w}(x)^T \mathbf{q}_w(t) \tag{6.4} $$

$$ \vartheta(x,t) = \boldsymbol{\vartheta}(x)^T \mathbf{q}_\vartheta(t). \tag{6.5} $$

The spatial shape functions have to fulfill the geometric boundary conditions, in this particular case for a clamped beam. For example, for the deflection $v(x,t)$ they can be chosen as

$$ v(x,t) = \begin{pmatrix} \frac{x^2}{L^2} - \frac{x^3}{L^3} \\ 3\frac{x^2}{L^2} - 2\frac{x^3}{L^3} \end{pmatrix}^T \begin{pmatrix} q_{v,1} \\ q_{v,2} \end{pmatrix}. \tag{6.6} $$

It turned out that two modes in each bending direction and one for the torsional direction are sufficient to reasonably describe the robot under consideration.

Therefore the vector of relative Ritz coefficients $\mathbf{q}_R$ (index R for Ritz) of one arm reads

$$\mathbf{q}_R^T = (q_{v,1} \;\; q_{v,2} \;\; q_{w,1} \;\; q_{w,2} \;\; q_{\vartheta}). \tag{6.7}$$

The absolute velocities of the mass element in the moving reference frame are

$$_R\mathbf{v}_{dm} = {_R\mathbf{v}_o} + {_R\widetilde{\boldsymbol{\omega}}_{IR}} \; {_R\mathbf{r}_{dm}} + {_R\dot{\mathbf{r}}_{dm}} \tag{6.8}$$

$$_R\boldsymbol{\omega}_{dm} = {_R\boldsymbol{\omega}_{IR}} + {_R\dot{\boldsymbol{\varphi}}_{dm}}, \tag{6.9}$$

where $_R\mathbf{v}_o$ is the translational velocity of the coupling point O and $_R\boldsymbol{\omega}_{IR}$ is the rotational velocity of the reference frame R. According to Fig. 6.2, this velocity can be further decomposed into guidance velocities $\boldsymbol{\omega}_F$, stemming from previous subsystems and the gear output coordinate $\dot{q}_A$, i.e.

$$_R\boldsymbol{\omega}_{IR} = {_R\boldsymbol{\omega}_F} + {_R\mathbf{e}_R}\dot{q}_A. \tag{6.10}$$

Therein $_R\mathbf{e}_R$ is the axis of rotation e.g. $_R\mathbf{e}_R^T = (0 \;\; 0 \;\; 1)$ for a rotation about the local z coordinate.

For a complete description of a subsystem i, describing velocities $\dot{\mathbf{y}}_i$ can be introduced. They have to contain the guidance velocities (translational $\mathbf{v}_o$, rotational $\boldsymbol{\omega}_F$) as well as the relative velocities, i.e. the motor velocity $\dot{q}_M$, link velocity $\dot{q}_A$ and Ritz coefficient velocities of the elastic beams $\dot{\mathbf{q}}_R$

$$\dot{\mathbf{y}}_n^T = (\mathbf{v}_o^T \;\; \boldsymbol{\omega}_F^T \;\; \dot{q}_M \;\; \dot{q}_A \;\; \dot{\mathbf{q}}_R^T). \tag{6.11}$$

6.2.2 *Multibody Systems with Open Chain Topology*

Using the concept of a kinematic chain, cf. Fig. 6.3, describing velocities $\dot{\mathbf{y}}_i$ of a considered subsystem i can be calculated by describing velocities of the predecessor system p multiplied by the kinematic transformation $\mathbf{T}_{ip}$ plus the relative parts of the considered subsystem.

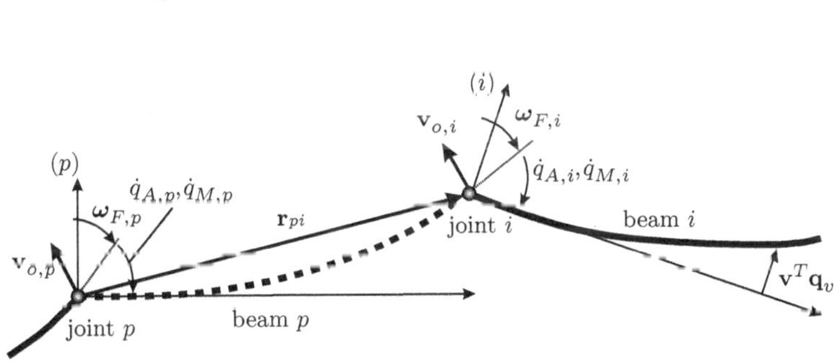

Fig. 6.3 Kinematic chain of flexible multibody systems

These relative parts can be separated into a local functional matrix $\mathbf{F}_i$ and relative minimal velocities $\dot{q}_M, \dot{q}_A, \dot{\mathbf{q}}_R$. Therefore the kinematic transformations for elastic subsystems as described in (6.11) is

$$\underbrace{\begin{pmatrix} \mathbf{v}_o \\ \boldsymbol{\omega}_F \\ \dot{q}_M \\ \dot{q}_A \\ \dot{\mathbf{q}}_R \end{pmatrix}_i}_{\dot{\mathbf{y}}_i} = \underbrace{\begin{bmatrix} \mathbf{R}_{ip} & \mathbf{R}_{ip}\tilde{\mathbf{r}}_{pi}^T & \mathbf{R}_{ip}\tilde{\mathbf{r}}_{pi}^T\mathbf{e}_R & 0 & \mathbf{R}_{ip}\frac{\partial \mathbf{r}_{pi}}{\partial \mathbf{q}_R} \\ 0 & \mathbf{R}_{ip} & \mathbf{R}_{ip}\mathbf{e}_R & 0 & \mathbf{R}_{ip}\frac{\partial \boldsymbol{\varphi}_{pi}}{\partial \mathbf{q}_R} \\ 0 & 0 & 0 & 0 & 0 \\ 0 & 0 & 0 & 0 & 0 \\ 0 & 0 & 0 & 0 & 0 \end{bmatrix}}_{\mathbf{T}_{ip}} \underbrace{\begin{pmatrix} \mathbf{v}_o \\ \boldsymbol{\omega}_F \\ \dot{q}_M \\ \dot{q}_A \\ \dot{\mathbf{q}}_R \end{pmatrix}_p}_{\dot{\mathbf{y}}_p} + \underbrace{\begin{bmatrix} 0 & 0 & 0 \\ 0 & 0 & 0 \\ 1 & 0 & 0 \\ 0 & 1 & 0 \\ 0 & 0 & \mathbf{I} \end{bmatrix}}_{\mathbf{F}_i} \underbrace{\begin{pmatrix} \dot{q}_M \\ \dot{q}_A \\ \dot{\mathbf{q}}_R \end{pmatrix}_i}_{\dot{\mathbf{q}}_i}, \tag{6.12}$$

see also [9] for other types of subsystems. With the Ritz ansatz, the connecting vectors between the subsystems can easily be written as

$$\mathbf{r}_{pi} = \mathbf{r}_{dm}(L, t), \boldsymbol{\varphi}_{pi} = \boldsymbol{\varphi}_{dm}(L, t). \tag{6.13}$$

The general formulation

$$\dot{\mathbf{y}}_i = \mathbf{T}_{ip}\dot{\mathbf{y}}_p + \mathbf{F}_i\dot{\mathbf{q}}_i \tag{6.14}$$

holds for all serial chained systems with tree structure. The kinematics of a multibody system with open chain topology consisting of N subsystems can be formulated by successively applying (6.14) as

$$\underbrace{\begin{pmatrix} \dot{\mathbf{y}}_1 \\ \dot{\mathbf{y}}_2 \\ \dot{\mathbf{y}}_3 \\ \vdots \\ \vdots \\ \dot{\mathbf{y}}_N \end{pmatrix}}_{\dot{\mathbf{y}}} = \underbrace{\begin{bmatrix} \mathbf{F}_1 & 0 & 0 & 0 & 0 \\ \mathbf{T}_{21}\mathbf{F}_1 & \mathbf{F}_2 & 0 & 0 & 0 \\ \mathbf{T}_{32}\mathbf{T}_{21}\mathbf{F}_1 & \mathbf{T}_{32}\mathbf{F}_2 & \mathbf{F}_3 & 0 & 0 \\ \vdots & \vdots & \ddots & 0 & 0 \\ \vdots & \vdots & \vdots & & 0 \\ \mathbf{T}_{N1}\mathbf{F}_1 & \mathbf{T}_{N2}\mathbf{F}_2 & \mathbf{T}_{N3}\mathbf{F}_3 & \cdots & \mathbf{F}_N \end{bmatrix}}_{\mathbf{F}} \underbrace{\begin{pmatrix} \dot{\mathbf{q}}_1 \\ \dot{\mathbf{q}}_2 \\ \dot{\mathbf{q}}_3 \\ \vdots \\ \vdots \\ \dot{\mathbf{q}}_N \end{pmatrix}}_{\dot{\mathbf{q}}}. \tag{6.15}$$

The transformation matrices $\mathbf{T}_{kj}$ for the serial chain are given by

$$\mathbf{T}_{kj} = \mathbf{T}_{k,p(k)} \times \mathbf{T}_{p(k),p(p(k))} \times \cdots \times \mathbf{T}_{s(j),j}, \tag{6.16}$$

wherein $p(k)$ denotes the predecessor of k and $s(j)$ denotes the successor of j. Therefore all the describing velocities $\dot{\mathbf{y}}$ can be calculated by the global functional matrix $\mathbf{F}$ applied to the vector of minimal velocities $\dot{\mathbf{q}}$. The global functional matrix $\mathbf{F}$ has lower triangular structure due the serial structure of the MBS. Details for tree structured systems can be found in [10].

6.3 Dynamical Modeling

6.3.1 *Equations of Motion*

The dynamical model of the FMBS is the basis for control design, trajectory optimization, parameter identification and therefore essential for the overall robot performance. The focus of this paper lies on articulated flexible link/joint robots as shown in Fig. 6.1. This robot consists of two types of subsystems. One consists of a motor, elastic gear and a rigid arm. This is quite simple and used for the first axis. The second subsystem is already introduced with Fig. 6.2. It consists of a motor, elastic gear, base body, elastic arm, tip body. Partitioning an MBS into subsystems simplifies the overall calculations. An efficient method to derive the EoM is the Projection Equation [5] in subsystem representation

$$\sum_{i=1}^{N} \left(\frac{\partial \dot{\mathbf{y}}_i}{\partial \dot{\mathbf{q}}}\right)^T (\mathbf{M}_i \ddot{\mathbf{y}}_i + \mathbf{G}_i \dot{\mathbf{y}}_i - \mathbf{Q}_i) = 0, \tag{6.17}$$

where $\mathbf{M}_i$, $\mathbf{G}_i$, $\mathbf{Q}_i$ are mass matrix, gyroscopic matrix (centrifugal, Coriolis) and generalized forces (friction, elastic potential, motor torques) of a subsystem, respectively. In the following the acceleration independent parts are abbreviated by

$$\mathbf{h}_i = \mathbf{G}_i \dot{\mathbf{y}}_i - \mathbf{Q}_i. \tag{6.18}$$

Each subsystem is associated with describing velocities $\dot{\mathbf{y}}_i$, and $\dot{\mathbf{q}}$ are the minimal velocities of the robot. The Jacobians of the describing velocities w.r.t. the minimal velocities project the subsystems EoM into the minimal space. The equations of motion in minimal representation are

$$\mathbf{M}(\mathbf{q})\ddot{\mathbf{q}} + \mathbf{h}(\mathbf{q}, \dot{\mathbf{q}}) = 0. \tag{6.19}$$

For the explicit calculation of the minimal accelerations, the global mass matrix has to be inverted which is computationally costly ($O(n^3)$), where n is the number of degrees of freedom. This can be avoided by using the $O(n)$ algorithm of Sect. 6.3.3. The main idea is to exploit the structure of the MBS in (6.17) in a Gaussian elimination process for which the lower triangular structure of the global functional matrix $\mathbf{F}$ from (6.15) is essential.

6.3.2 *Subsystems*

Figure 6.2 shows the elastic subsystem for the robot under consideration. The rigid subsystem is a special case of the elastic subsystem. For the calculation of the corresponding subsystem matrices, the velocities of the center of gravity (rigid bodies) and the cross section dm (elastic bodies) have to be written as function of the describing velocities of the subsystem

$$\begin{pmatrix} \mathbf{v}_{dm} \\ \boldsymbol{\omega}_{dm} \end{pmatrix} = \overline{\mathbf{F}}_i^T \underbrace{(\mathbf{v}_o^T \; \boldsymbol{\omega}_F^T \; \dot{q}_M \; \dot{q}_A \; \dot{\mathbf{q}}_R^T)^T}_{\dot{\mathbf{y}}_i} . \tag{6.20}$$

The Jacobian $\overline{\mathbf{F}}_i^T$ projects these velocities to the space of describing velocities $\dot{\mathbf{y}}_i$. Thus, the mass matrix for the subsystem in Fig. 6.2 reads

$$\mathbf{M}_i = \sum_{j=1}^{N_{rb}} (\overline{\mathbf{F}}^T \overline{\mathbf{M}} \overline{\mathbf{F}})_j + \sum_{j=1}^{N_{el}} \left(\int_o^L (\overline{\mathbf{F}}^T \frac{d\overline{\mathbf{M}}}{dx} \overline{\mathbf{F}}^T) dx \right)_j \tag{6.21}$$

that is composed of N_{rb} rigid bodies and N_{el} elastic bodies. $\overline{\mathbf{M}}$ and $d\overline{\mathbf{M}}$ are the mass matrices of a single rigid body and the cross section , respectively. For the calculation of the remaining subsystem matrices we refer to [9]. Note, also the effect of dynamical stiffening can be effectively included into the subsystem terms, see [12] for details.

6.3.3 Efficient Evaluation of Minimal Accelerations

Details on the derivation of the efficient evaluation of the minimal accelerations by the $O(n)$ algorithm can be found in [11]. Below, only the algorithm is reproduced.

Algorithm 1

1. Step: Calculate

$$\mathbf{T}_{ip}, \dot{\mathbf{T}}_{ip}, \dot{\mathbf{y}}_i \tag{6.22}$$

for $i = 1 \ldots N$.

2. Step: Determine

$$\mathbf{M}_p^* = \mathbf{M}_p + \mathbf{T}_{ip}^{*T} \mathbf{M}_i^* \mathbf{T}_{ip} \tag{6.23}$$

$$\mathbf{h}_p^* = \mathbf{h}_p + \mathbf{T}_{ip}^{*T} \left(\mathbf{M}_i^* \dot{\mathbf{T}}_{ip} \dot{\mathbf{y}}_p + \mathbf{h}_i^* \right) \tag{6.24}$$

$$\mathbf{T}_{ip}^* = \mathbf{N}_i^T \mathbf{T}_{ip}, \quad \mathbf{N}_i = \left[\mathbf{I} - \mathbf{M}_i^* \mathbf{J}_i \right]$$

$$\mathbf{J}_i := \mathbf{F}_i \mathbf{M}_{Ri}^{-1} \mathbf{F}_i^T, \; \mathbf{M}_{Ri} := \left[\mathbf{F}_i^T \mathbf{M}_i^* \mathbf{F}_i \right]$$

for $p = N \ldots 1$.

3. Step: Solve

$$\ddot{\mathbf{q}}_i = -\mathbf{M}_{Ri}^{-1} \mathbf{F}_i^T \left\{ \mathbf{M}_i^* \left(\mathbf{T}_{ip} \ddot{\mathbf{y}}_p + \dot{\mathbf{T}}_{ip} \dot{\mathbf{y}}_p \right) + \mathbf{h}_i^* \right\} \tag{6.25}$$

$$\ddot{\mathbf{y}}_i = \mathbf{T}_{ip} \ddot{\mathbf{y}}_p + \dot{\mathbf{T}}_{ip} \dot{\mathbf{y}}_p + \mathbf{F}_i \ddot{\mathbf{q}}_i \tag{6.26}$$

for $i = 1 \ldots N$.

Note, only the reduced mass matrix $\mathbf{M}_{Ri}$ must be inverted. This matrix has dimension according to the relative DoF of the ith subsystem. In contrast to other $O(n)$ formulations mentioned in Sect. 6.1, with the above algorithm no distinction between rigid or elastic bodies has to be made.

6.4 Contact Simulation

6.4.1 EOM in Closed from in Terms of Minimal Coordinates

If the robotic system enters an additional contact with the environment, a kinematic loop $\boldsymbol{\phi} = \mathbf{r}_E - \mathbf{r}_C = 0$ is closed. Therein $\boldsymbol{\phi}$ is the geometric constraint and $\mathbf{r}_E$, $\mathbf{r}_C$ are position vectors to the end-effector of the robot and to the contact point on the surface, respectively. Figure 6.4 on the left shows a scenario for the system without contact and on the right, the system where the kinematic loop is closed.

In this case, the EoM (6.19) must be completed by the generalized constraint reactions

$$\mathbf{M}(\mathbf{q})\ddot{\mathbf{q}} + \mathbf{h}(\mathbf{q}, \dot{\mathbf{q}}) - \mathbf{J}^T\boldsymbol{\lambda} = 0, \tag{6.27}$$

where $\mathbf{J} \in \mathbb{R}^{n,N_c}$ is the constraint Jacobian and $\boldsymbol{\lambda} \in \mathbb{R}^{N_c}$ are Lagrange multipliers that represent the contact forces. In this case, a number of N_c geometric constraints $\boldsymbol{\phi} = 0$ can be active. This constraints also holds for the velocity and acceleration

$$\dot{\boldsymbol{\phi}} = \left(\frac{\partial \boldsymbol{\phi}}{\partial \mathbf{q}}\right)\dot{\mathbf{q}} = \left(\frac{\partial \dot{\boldsymbol{\phi}}}{\partial \dot{\mathbf{q}}}\right)\dot{\mathbf{q}} := \mathbf{J}\dot{\mathbf{q}} \tag{6.28}$$

$$\ddot{\boldsymbol{\phi}} = \mathbf{J}\ddot{\mathbf{q}} + \dot{\mathbf{J}}\dot{\mathbf{q}} = 0. \tag{6.29}$$

The above constraints equations are valid as long as the contact is closed.

Obviously, the kth column of $\mathbf{J}^T$ corresponds to the component λ_k of the contact force. Therefore, the Jacobian is written as

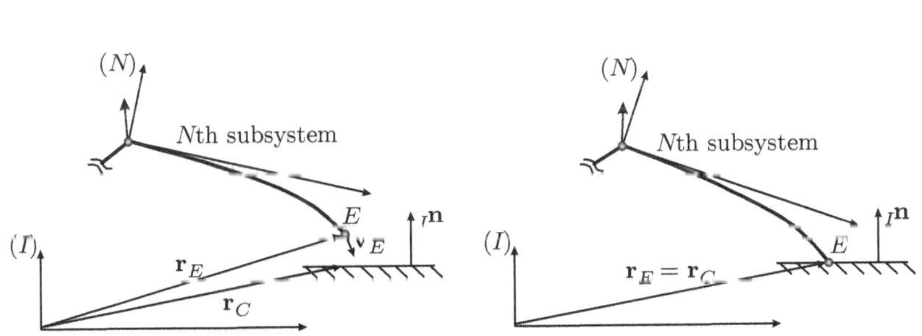

Fig. 6.4 Contact detail; left: before contact; right: in contact

$$\mathbf{J}^T = [\mathbf{J}_1^T, \ldots, \mathbf{J}_k^T, \ldots, \mathbf{J}_{N_c}^T]. \tag{6.30}$$

This is important for the subsequent recursive formulation in Sect. 6.4.2. Combining (6.27) and (6.29) leads to the differential-algebraic system

$$\begin{bmatrix} \mathbf{M} & -\mathbf{J}^T \\ -\mathbf{J} & 0 \end{bmatrix} \begin{pmatrix} \ddot{\mathbf{q}} \\ \boldsymbol{\lambda} \end{pmatrix} = \begin{pmatrix} -\mathbf{h} \\ \dot{\mathbf{J}}\dot{\mathbf{q}} \end{pmatrix} \tag{6.31}$$

that can be solved for the accelerations $\ddot{\mathbf{q}}$ and $\boldsymbol{\lambda}$. From a simulation point of view it can be advantageous to calculate $\boldsymbol{\lambda}$ and $\ddot{\mathbf{q}}$ in a staggered approach. Reformulating (6.31) yields

$$\boldsymbol{\lambda} = \left[\mathbf{J}\mathbf{M}^{-1}\mathbf{J}^T\right]^{-1} \left(\mathbf{J}\mathbf{M}^{-1}\mathbf{h} - \dot{\mathbf{J}}\dot{\mathbf{q}}\right). \tag{6.32}$$

which can then be used in (6.27) to solve for the accelerations

$$\ddot{\mathbf{q}} = -\mathbf{M}^{-1}(\mathbf{h} - \mathbf{J}^T\boldsymbol{\lambda}). \tag{6.33}$$

In (6.33) and (6.32) the inverse of the overall mass matrix $\mathbf{M}^{-1}$ is needed. The computation time of this inversion increases with $O(n^3)$ and is therefore computationally expensive.

A transition of the closed system to the open system is indicated by a changing sign of the contact force. In this case $\boldsymbol{\lambda}$ is set to zero and the calculations start again.

Note, since only the accelerations of the constraints enter the calculations with (6.29) in a succeeding time integration of the accelerations a numerical drift will occur. This can easily be handled with Baumgarte's constraint stabilization [2] and does not change the algorithms. However, this is not topic of the present contribution.

The transition between the free system and constrained system is characterized by an impact. During this impact the generalized momentum is conserved. Integrating (6.33) over the very short impact time interval $[t_b, t_e]$ yields

$$\int_{t_b}^{t_e} (\mathbf{M}(\mathbf{q})\ddot{\mathbf{q}} + \mathbf{h}(\mathbf{q}, \dot{\mathbf{q}}) - \mathbf{J}_n(\mathbf{q})^T\lambda_n)dt = \mathbf{M}(\mathbf{q})\ (\dot{\mathbf{q}}_e - \dot{\mathbf{q}}_b) - \mathbf{J}_n(\mathbf{q})^T\Lambda_n = 0. \tag{6.34}$$

This result follows with the assumption that the position, i.e. $\mathbf{q}$, does not change during the impact and that the integral of the generalized forces $\mathbf{Q}$ in (6.18) are small compared to the impulsive impact forces. In (6.34) subscript b and e indicate the begin and end of the impact, respectively. $\mathbf{J}_n$ is the constraint Jacobian in normal direction (n) and Λ_n is the impulsive force in normal direction

$$\Lambda_n = \int_{t_b}^{t_e} \lambda_n dt. \tag{6.35}$$

Using Newton's impact law

$$\dot{\phi}_{n,e} = -\epsilon \dot{\phi}_{n,b} \tag{6.36}$$

means that the normal velocity at the end of the impact $\dot{\phi}_{n,e}$ is equal to the negative normal velocity $\dot{\phi}_{n,b}$ at the begin, damped by the restitution coefficient ϵ [18]. With (6.34), and (6.36) the impulsive force yields

$$\Lambda_n = -(\mathbf{J}_n \mathbf{M}^{-1} \mathbf{J}_n^T)^{-1}(1+\epsilon)\mathbf{J}_n \dot{\mathbf{q}}_b \tag{6.37}$$

Therefore the velocities $\dot{\mathbf{q}}_e$ after the impact can be determined with (6.34) as

$$\dot{\mathbf{q}}_e = \dot{\mathbf{q}}_b - \mathbf{M}^{-1}\mathbf{J}_n^T(\mathbf{J}_n \mathbf{M}^{-1} \mathbf{J}_n^T)^{-1}(1+\epsilon)\mathbf{J}_n \dot{\mathbf{q}}_b. \tag{6.38}$$

These velocities contain again $\mathbf{M}^{-1}$. Note, for integration of the system in a simulation, the solver has to be restarted with these new velocities.

6.4.2 *Recursive Formulation and $O(n)$ Algorithm*

As mentioned in the last section, the inversion of the overall mass matrix is computationally expensive leading to long simulation times. However, (6.32) gives rise to $O(n)$-algorithms to solve for the generalized accelerations, which are faster than the direct $O(n^3)$ resolution for large n.

6.4.2.1 Idea

The idea is to separate specific terms in (6.32) and calculate them by partially evaluating Algorithm 1. The first step is the forward/backward/forward recursion to calculate accelerations

$$\ddot{\mathbf{q}}_{[0]} := -\mathbf{M}^{-1}\mathbf{h} \tag{6.39}$$

of the unconstrained system. A subscript in square bracket indicates the number of $O(n)$ algorithm calls that are used. Since the $O(n)$ algorithm is applied, the inversion of the mass matrix is avoided. This accelerations are part of $\boldsymbol{\lambda}$. Avoiding the calculation of $\mathbf{M}^{-1}$ in the first part of (6.32) $\mathbf{J}\mathbf{M}^{-1}\mathbf{J}^T$ is a little bit tricky. As already mentioned in (6.30) the constraint Jacobian can be separated into its columns where each column is associated with a specific geometric constraint and a contact force in the specific direction. In a second step the first constraint $\mathbf{J}_1$ is considered. We can introduce amended generalized forces as $\mathbf{h}_{[1]} := \mathbf{h} - \mathbf{J}_1^T$ and determine auxiliary accelerations $\ddot{\mathbf{q}}_{[1]} = -\mathbf{M}^{-1}\mathbf{h}_{[1]}$ using an $O(n)$ method.

The difference $-\ddot{\mathbf{q}}_{[0]} + \ddot{\mathbf{q}}_{[1]}$ is then identical to the first column of $\mathbf{M}^{-1}\mathbf{J}^T$. This is repeated for all N_c constraints by defining

$$\mathbf{h}_{[k]} := \mathbf{h} - \mathbf{J}_k^T \tag{6.40}$$

and solving for auxiliary accelerations

$$\ddot{\mathbf{q}}_{[k]} := -\mathbf{M}^{-1}(\mathbf{h} - \mathbf{J}_k^T), k = 1 \dots N_c. \tag{6.41}$$

The difference $-\ddot{\mathbf{q}}_{[0]} + \ddot{\mathbf{q}}_{[k]}$ delivers the kth column of $\mathbf{M}^{-1}\mathbf{J}^T$. Thus, the Eq. (6.32) can be rewritten to determine $\boldsymbol{\lambda}$ as

$$\boldsymbol{\lambda} = \left[\mathbf{J}[-\ddot{\mathbf{q}}_{[0]} + \ddot{\mathbf{q}}_{[1]}, \dots, -\ddot{\mathbf{q}}_{[0]} + \ddot{\mathbf{q}}_{[k]}, \dots, -\ddot{\mathbf{q}}_{[0]} + \ddot{\mathbf{q}}_{[N_c]}]\right]^{-1} \left(\mathbf{J}(-\ddot{\mathbf{q}}_{[0]}) - \dot{\mathbf{J}}\dot{\mathbf{q}}\right). \tag{6.42}$$

In a final step the accelerations are calculated from the known contact forces by $\ddot{\mathbf{q}} = -\mathbf{M}^{-1}(\mathbf{h} - \mathbf{J}^T\boldsymbol{\lambda})$ using an $O(n)$ forward recursion.

6.4.2.2 Implementation

As indicated by (6.40), the vector of generalized forces $\mathbf{h}$ is augmented by the corresponding column of $\mathbf{J}$. Due to our subsystem formulation and the $O(n)$ algorithm in Sect. 6.3.3, $\mathbf{h}$ is not directly available. In our formulation, only subsystem matrices $\mathbf{M}_i, \mathbf{h}_i$ can be used for calculations. Typically, contact arises at the end-effector E of the robot, where the corresponding forces acting on the Nth subsystem are $\mathbf{h}_E$. The columns of the Jacobian $\mathbf{J}$ (6.30) are associated to the generalized velocities $\dot{\mathbf{q}}$ (whole model). However, for the contact modeling using the subsystem formulation also the constraint Jacobian has to be defined on subsystem level. The corresponding describing velocities are $\dot{\mathbf{y}}_N$ (of the last subsystem at which the end-effector is attached). The Jacobian can be split according to the subsystem formulation

$$\mathbf{J} = \left(\frac{\partial \dot{\boldsymbol{\phi}}}{\partial \dot{\mathbf{q}}}\right) = \left(\frac{\partial \dot{\boldsymbol{\phi}}}{\partial \dot{\mathbf{y}}_N}\right)\left(\frac{\partial \dot{\mathbf{y}}_N}{\partial \dot{\mathbf{q}}}\right). \tag{6.43}$$

The first term is the transformation of the describing velocities of the contact subsystem to the constraint velocities followed by the transformation of the velocities $\dot{\mathbf{q}}$ to $\dot{\mathbf{y}}_N$.

In the robotic application we only consider translational contact constraints. Typically, the geometric constraints $\boldsymbol{\phi}$ and thus its time derivative $\dot{\boldsymbol{\phi}}$ are defined in the world frame (I). Choosing an appropriate constraint direction k one gets

$${}_I\dot{\phi}_k = {}_I\mathbf{n}_k{}^T\, {}_I\mathbf{v}_E, \tag{6.44}$$

see again Fig. 6.4, where $\mathbf{n}_k$ is the normal vector. For the subsystem formulation, a transformation to the Nth subsystem frame is facilitated using the rotation matrix $\mathbf{R}_{NI}$, i.e.

$$ {}_I\dot{\phi}_k = {}_I\mathbf{n}_k{}^T \underbrace{\mathbf{R}_{NI}^T \mathbf{R}_{NI}}_{\mathbf{I}} {}_I\mathbf{v}_E = {}_N\mathbf{n}_k{}^T {}_N\mathbf{v}_E, \tag{6.45}$$

and the corresponding Jacobian for the contact subsystem, cf. (6.43), is

$$\left(\frac{\partial_I \dot{\boldsymbol{\phi}}}{\partial \dot{\mathbf{y}}_N}\right) = {}_N\mathbf{n}^T \left(\frac{\partial \mathbf{v}_E}{\partial \dot{\mathbf{y}}_N}\right). \tag{6.46}$$

The last term is the kinematic mapping of end-effector velocities ${}_N\dot{\mathbf{v}}_E$ to the describing velocities $\dot{\mathbf{y}}_N$ of the last subsystem. With this Jacobian the subsystem forces $\mathbf{h}_N$ (see (6.24)) of the contact subsystem N are augmented to

$$\mathbf{h}_{N,[k]} := \mathbf{h}_N - \left(\frac{\partial \dot{\phi}_k}{\partial \dot{\mathbf{y}}_N}\right)^T \tag{6.47}$$

in analogy to (6.18). Then parts of the $O(n)$ Algorithm are performed to get $\ddot{\mathbf{q}}_{[k]}$. Since only $\mathbf{h}_N$ is changed, while the mass matrices $\mathbf{M}_i$ and kinematics $\dot{\mathbf{y}}_i$ remain unchanged, only step 2 and step 3 of the $O(n)$ algorithm without adaptation of the mass matrix have to be performed. Therefore all terms in (6.42) can be evaluated and $\boldsymbol{\lambda}$ is used in a last step to calculate $\ddot{\mathbf{q}}$ by setting

$$\mathbf{h}_N := \mathbf{h}_N - \left(\frac{\partial \dot{\boldsymbol{\phi}}}{\partial \dot{\mathbf{y}}_N}\right)^T \boldsymbol{\lambda} \tag{6.48}$$

again and partially evaluating the $O(n)$ algorithm (Step 2 and Step 3 without mass matrices transformations). The following algorithm can be formulated:

The last topic that is addressed within the theoretical part of this paper is impact. By observing (6.37) and (6.38) where $\mathbf{M}^{-1}$ is needed, it becomes obvious that it appears only in the combination with the contact Jacobian, i.e. $\mathbf{J}_n^T$. Therefore these entries can analogously be calculated as above. An algorithm for this is

6.5 Simulation Results

The simulation is performed in Matlab/Simulink. Figure 6.5 shows the principle simulation scheme. The trajectory generation plans desired trajectories $\mathbf{r}_{E,d}$ for the end-effector of the robot. These are transformed to desired motor positions $\mathbf{q}_{M,d}$ via inverse kinematics calculations, see e.g. [20]. The error between desired and

Algorithm 2

1. Step: Calculate $-\ddot{\mathbf{q}}_{[0]}$ of the unconstrained system by applying Algorithm 1 (unconstrained system).
2. Step: For all $k = 1 \dots N_c$ constraints, compute

$$\mathbf{h}_{N,[k]} := \mathbf{h}_N - \left(\frac{\partial \dot{\phi}_k}{\partial \dot{\mathbf{y}}_N} \right)^T. \tag{6.49}$$

   Apply step 2 and step 3 of Algorithm 1 with new vector $\mathbf{h}_{N,[k]}$ to calculate corresponding accelerations $\ddot{\mathbf{q}}_{[k]}$.
3. Step: Combine the accelerations to the corresponding columns of

$$\mathbf{M}^{-1}\mathbf{J}^T = [\mathbf{M}^{-1}\mathbf{J}_1^T, \dots, \mathbf{M}^{-1}\mathbf{J}_k^T, \dots] := [-\ddot{\mathbf{q}}_{[0]} + \ddot{\mathbf{q}}_{[1]}, \dots, -\ddot{\mathbf{q}}_{[0]} + \ddot{\mathbf{q}}_{[k]}, \dots]. \tag{6.50}$$

4. Step: Evaluate $\boldsymbol{\lambda}$ with (6.42)
5. Step: Apply step 2 and step 3 of Algorithm 1 with new vector

$$\mathbf{h}_{N,\lambda} := \mathbf{h}_N - \left(\frac{\partial \dot{\boldsymbol{\phi}}}{\partial \dot{\mathbf{y}}_N} \right)^T \boldsymbol{\lambda} \tag{6.51}$$

   and calculate accelerations $\ddot{\mathbf{q}}$.

Algorithm 3

1. Step: Calculate $-\ddot{\mathbf{q}}_{[0]}$ of the unconstrained system by applying Algorithm 1 (unconstrained system).
2. Step: Add

$$\mathbf{h}_{N,n} := \mathbf{h}_N - \left(\frac{\partial \dot{\phi}_n}{\partial \dot{\mathbf{y}}_N} \right)^T \tag{6.52}$$

   Apply step 2 and step 3 of Algorithm 1 with new vector $\mathbf{h}_{N,n}$ and calculate corresponding accelerations $\ddot{\mathbf{q}}_{[n]}$.
3. Step: Combine the accelerations to the corresponding column of

$$\mathbf{M}^{-1}\mathbf{J}_n^T := [-\ddot{\mathbf{q}}_{[0]} + \ddot{\mathbf{q}}_{[n]}] \tag{6.53}$$

4. Step: Evaluate $\dot{\mathbf{q}}_e$ with (6.32)

$$\dot{\mathbf{q}}_e = \dot{\mathbf{q}}_b - [-\ddot{\mathbf{q}}_{[0]} + \ddot{\mathbf{q}}_{[n]}](\mathbf{J}_n[-\ddot{\mathbf{q}}_{[0]} + \ddot{\mathbf{q}}_{[n]}])^{-1}(1+\epsilon)\mathbf{J}_n\dot{\mathbf{q}}_b. \tag{6.54}$$

5. Step: Restart integration solver with new state vector $\mathbf{x}^T = (\mathbf{q}^T, \dot{\mathbf{q}}_e^T)$.

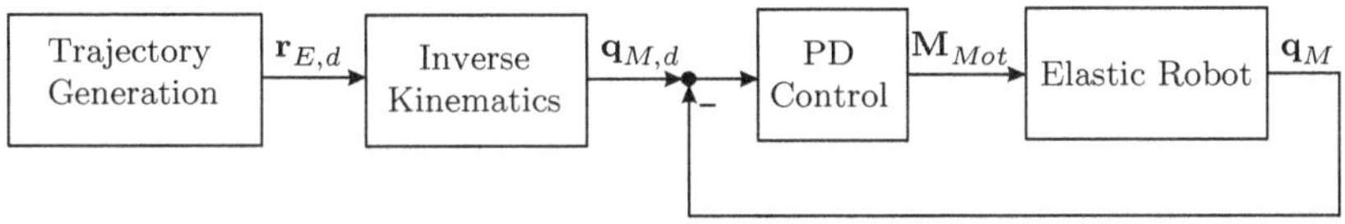

Fig. 6.5 Simulation scheme

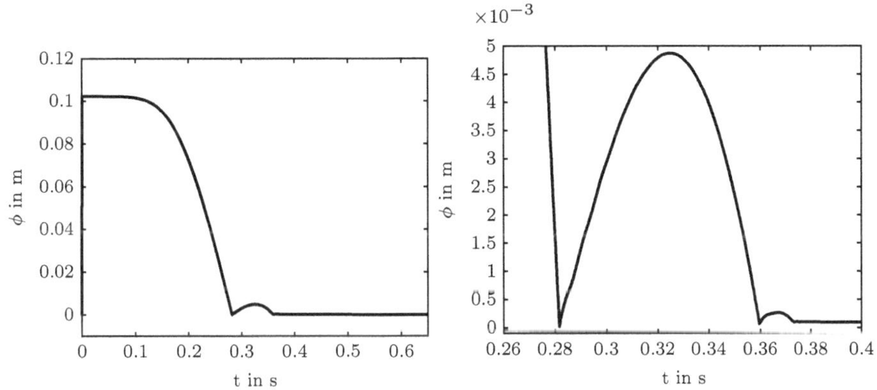

Fig. 6.6 Geometric constraint ϕ; left: full simulation; right: zoom

simulated motor positions are used in a standard proportional- differential (PD) control law to generate corresponding motor torques $\mathbf{M}_{Mot}$. Since the focus of this paper is on the (contact) simulation of the FMBS, no vibration damping control is included. The reader is referred to [20] for this topic. The $O(n)$ algorithms from Sects. 6.3 and 6.4 are implemented in C++ using the Eigen library [7] and is included into the simulation environment as an S-function.

To verify the algorithms, the following example is considered. The robot starts in a position without contact, cf. Fig. 6.1. A trajectory generator plans a trajectory for the end-effector in negative z-direction of $\Delta z = -0.15$ m. Only axes 2 and 3 are used for this movement. Additionally, only one contact in z-direction is considered in this example leading to a single contact force. The corresponding values of the geometric constraint function $\phi = {}_I\mathbf{r}_{E,z} - {}_I\mathbf{r}_{C,z}$ is shown in Fig. 6.6 for the full simulation (left) and a zoom of the impact/contact phase (right). In this example, no force is applied in the lateral directions x, y. Typically, friction forces that depend on the normal force are included in these directions.

The impact occurs at a distance of about $\Delta z = -0.1$ m and the end-effector lifts off until the motor control moves it back and again an impact occurs. This happens several times until the contact velocity is below a threshold (here 0.01 m/s) the contact is assumed to be closed and the topology of the system is changed. In this phase, the contact forces are calculated.

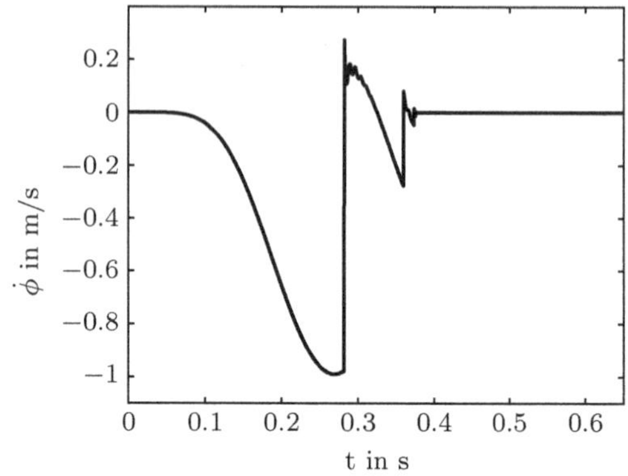

Fig. 6.7 Constraint velocity $\dot{\phi}$

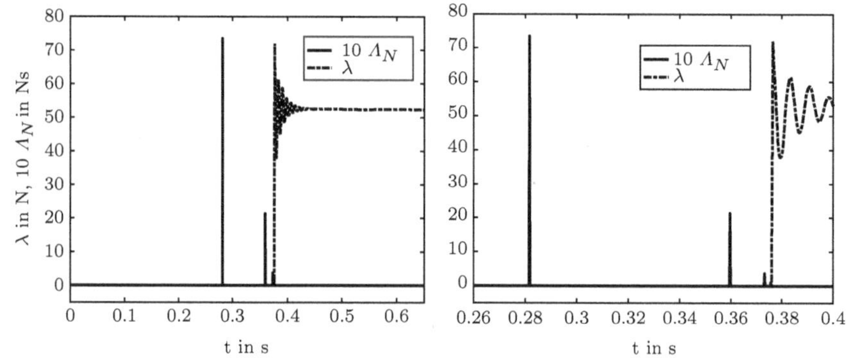

Fig. 6.8 Scaled impulsive force Λ_N and contact force λ; left: full simulation; right: zoom

Figure 6.7 depicts the corresponding constraint velocity $\dot{\phi}$. The impact velocity between the robots end-effector and the virtual force sensor is about 1 m/s at time 0.28 s. A restitution coefficient of $\epsilon = 0.3$ is chosen as can be seen in the figure. At time 0.36 s a second impact occurs. After the third impact, the normal velocity is sufficiently small that the topology is changed to contact. The corresponding contact force can be seen in Fig. 6.8. This force can be measured by a force sensor in a real application. Due to the impact modeling, the first peaks are the impulsive forces scaled by factor 10. When the topology is changed, only contact forces can be seen. Due to the impact the system starts to vibrate with about 140 Hz which corresponds to a higher bending eigenfrequency of the system. The motor torques for the present simulation example are shown in Fig. 6.9. The end-effector moves down and at time 0.19 s the deceleration phase starts indicated by the increasing torques. The impact at time 0.28 s is clearly visible by torque peaks (reactions from motor position control). Due to the contact force, the steady state torques in the contact phase are lower than they are for the free system.

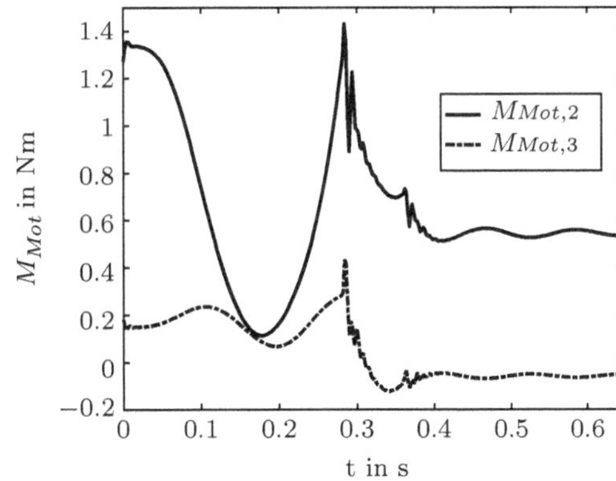

Fig. 6.9 Motor torques

6.6 Conclusion

In this paper, the dynamical modeling and calculation of accelerations for flexible multibody systems using an $O(n)$ method, i.e. without inverting the overall mass matrix, is presented. Recently such robotic systems are often used for grinding, deburring etc. which means that the robot goes in contact with the environment. From a simulation point of view this can be tackled including corresponding contact forces. Typically these contact forces depend again on the inverse of the overall mass matrix. An efficient algorithm for avoiding this inversion is presented where the accelerations can again be calculated by applying steps of the $O(n)$ algorithm. The proposed algorithm can also be applied to closed loop kinematics like parallel kinematic machines. Special attention is also given to the transition phase between the free system and the system in contact. This is done with the help of Newton's impact law where the velocities change at the impact regarding a restitution coefficient. With the new algorithm these velocities can also be calculated without inversion of the overall mass matrix.

A simulation example has shown the effectiveness of the proposed methods. An elastic robot is controlled such that the end-effector goes in fast contact with a stiff environment (steel – steel contact). The results are presented for a motor-position controlled robot without any vibration damping. Future investigations include simulations with a vibration damped robot, where the vibration damping is based on acceleration measurements. Also an experimental verification has to be done with the real robot. Additional measurements with a laser tracker especially in the impact phase will give new insights.

Acknowledgements This work has been supported by the Austrian COMET-K2 program of the Linz Center of Mechatronics (LCM).

References

1. Attia, H.A.: A recursive method for the dynamic analysis of mechanical systems in spatial motion. Acta Mech. **167**, 41–55 (2004)
2. Baumgarte, J.: Stabilization of constraints and integrals of motion in dynamical systems. Comput. Methods Appl. Mech. Eng. **1**, 490–501 (1972)
3. Brandl, H., Johanni, R., Otter, M.: A very efficient algorithm for the simulation of robots and similar multibody systems without inversion of the mass matrix. In: Proceedings of the IFAC International Symposium on Theory of Robots, Viennam, pp. 365–370 (1986)
4. Brandl, H., Johanni, R., Otter, M.: An algorithm for the simulation of multibody systems with kinematic loops. In: Proceedings of 7th IFToMM World Congress on the Theory of Machines and Mechanisms, Sevilla, pp. 407–411 (1987)
5. Bremer, H.: Elastic Multibody Dynamics: A Direct Ritz Approach. Springer, Heidelberg (2008)
6. Dwivedy, S.K., Eberhard, P.: Dynamic analysis of flexible manipulators, a literature review. Mech. Mach. Theory **41**(7), 749–777 (2006)
7. Eigen C++ Library http://eigen.tuxfamily.org
8. Featherstone, R.: Rigid Body Dynamics Algorithms. Springer, New York (2008)
9. Gattringer, H.: Starr-elastische Robotersysteme (Theorie und Anwendungen). Springer, Heidelberg (2011)
10. Gattringer, H., Müller, A.: Dynamic formulations and computational algorithms. In: Goswami, A., Vadakkepat, P. (eds.) Humanoid Robotics: A Reference. Springer, Netherlands (2017)
11. Gattringer, H., Bremer, H., Kastner, M.: Efficient dynamic modeling for rigid multibody systems with contact and impact. Acta Mech. **219**, 111–128 (2011)
12. Gattringer, H., Müller, A., Springer, K., Jörgl, M.: An efficient method for the dynamical modeling of serial elastic link/joint robots. In: Moreno-Diaz, R., et al. (eds.) Computer Aided Systems Theory – EUROCAST 2015. Series Lecture Notes in Computer Science, vol. 9520, pp. 689–697. Springer, Heidelberg (2015)
13. Glocker, C.: Set-Valued Force Laws – Dynamics of Non-smooth Systems. Springer, Berlin/Heidelberg (2001)
14. Höbarth, W., Gattringer, H., Bremer, H.: Modeling and control of an articulated robot with flexible links/joints. In: Proceedings of the 9th International Conference on Motion and Vibration Control, Garching (2008)
15. Khalil, W.: Dynamic modeling robots using recursive Newton-Euler techniques. In: Filipe, J., Cetto, J., Ferrier, J. (eds.) Proceedings of the 7th International Conference on Informatics in Control, Automation and Robotics, vol. 2, pp. 19–31. SciTePress, Portugal (2010)
16. Luh, J.Y.S., Walker, M.W., Paul, R.P.C.: On-line computational scheme for mechanical manipulators. ASME J. Dyn. Syst. Meas. Control **102**, 69–76 (1980)
17. Mohan, A., Saha, S.K.: A recursive, numerically stable, and efficient simulation algorithm for serial robots. Multibody Syst. Dyn. **17**, 291–319 (2007)
18. Pfeiffer, F.: Mechanical System Dynamics. Springer, Berlin (2008)
19. Siciliano, B., Khatib, O.: Springer Handbook of Robotics. Springer, Cham (2016)
20. Staufer, P., Gattringer, H.: State estimation on flexible robots using accelerometers and angular rate sensors. Mechatronics **22**, 1042–1049 (2012)
21. Walker, M.W., Orin, D.E.: Efficient dynamic computer simulation of robotic mechanisms. ASME J. Dyn. Syst. Meas. Control **104**, 205–211 (1982)

Chapter 7
Developing a 3-D, Lumped-Mass Model to Present Behaviour of Large Deformation Surface Based Continuum Robots

Hossein Habibi, Rongjie Kang, Ian D. Walker, Isuru S. Godage, and David T. Branson

Abstract The deployment of continuum robotic surfaces has strong potential over a wide range of engineering disciplines. To allow such compliant, actively actuated surfaces to be controlled accurately and efficiently, reliable kinematic and dynamic models are required. The main challenge appears when the continuum surfaces become very flexible and undergo large deformations, an issue which is little studied in continuum robotics to date. This paper tackles this problem through the application of a lumped-mass approach for analysis of continuum surfaces that are subject to large deformations due to either gravity or external loading applied by representative flexible actuators. The developed model describes the surface kinematics by providing a means of solving for the displacement profile across the surface. The model takes into account all the essential factors such as gravitational effects, material properties of a flexible plate, inertial forces, material damping, and

---

H. Habibi (✉)
School of Science, Engineering and Design, Teesside University, Middlesbrough, UK
e-mail: H.Habibi@tees.ac.uk

R. Kang
Key Laboratory of Mechanism Theory and Equipment Design of Ministry of Education, School of Mechanical Engineering, Tianjin University, Tianjin, China
e-mail: rjkang@tju.edu.cn

I. D. Walker
Department of Electrical & Computer Engineering, Clemson University, Clemson, SC, USA
e-mail: iwalker@clemson.edu

I. S. Godage
School of Computing, DePaul University, Chicago, IL, USA
e-mail: igodage@depaul.edu

D. T. Branson
Advanced Manufacturing Technology Research Group, Faculty of Engineering, University of Nottingham, Nottingham, UK
e-mail: David.Branson@nottingham.ac.uk

E. Zahariev, J. Cuadrado (eds.), *IUTAM Symposium on Intelligent Multibody Systems – Dynamics, Control, Simulation*, IUTAM Bookseries 33,
https://doi.org/10.1007/978-3-030-00527-6_7

in-depth shear effects across the surface. An experimental setup has been developed to test an actuated flexible surface under different boundary conditions, with results showing mean percentage error of 4.8% at measured surface points.

Keywords Large deformation continuum surfaces · Soft robotics · Lumped-mass modelling

7.1 Introduction

Surfaces capable of large deformations are progressively being considered to improve performance across a wide range of engineering applications. This can range from manufacturing processes such as composites formation or devising nature-inspired surface-like robots. Such flexible surfaces undergoing large deformations could also be used in airplane body structures to achieve better aerodynamic lift control, or used in vehicles to reduce drag. In terms of healthcare such soft systems are already in use [1] with controlled surfaces for use as exoskeletons under development by multiple research groups.

To employ such Large Displacement Continuum Surfaces (LDCS) efficiently, they need to be well adapted to the use of smart, controlled actuation that deforms the surfaces into an accurate, desired profile. In fact, the surface can be considered as a novel end-effector of a control system which is viewed and analysed within the area of continuum robotics. Figure 7.1 depicts a schematic LDCS deformed into multiple curvatures by means of actuation elements mounted on the surface edges.

In the area of continuum robotics, arms and manipulators have been well studied. Examples can be found in [2–9] where most of these are achieved through nature-inspired motion to be utilised for different tasks in industry, healthcare, manufacturing, or environments. However, unlike the relatively prolific number of studies on arm based structures, the research on LDCS within the continuum

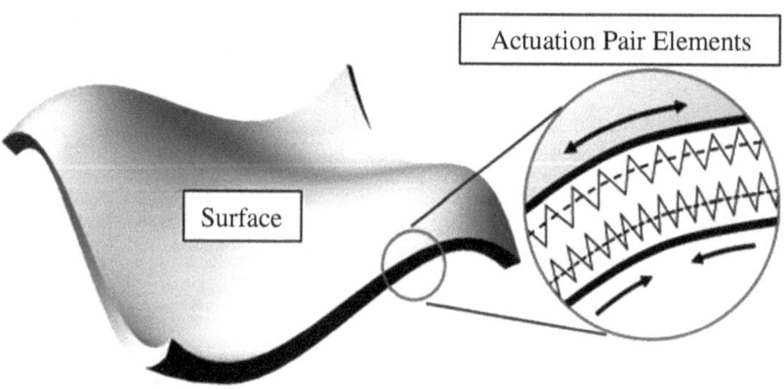

Fig. 7.1 Schematic diagram of a large deformation continuum surfaces (LDCS). Actuation elements in this case are placed along the edge, resulting in localized curvature through antagonistic actions

robotics area is relatively limited where to date LDCS have largely operated in 'open loop' with the placement and control of actuation elements based mainly on user intuition and experience. So far there is limited work attempting to provide models for such actuated continuum surfaces to enable model based control and simulation. This results in largely 'trial and error' based methods in their design and control that increases production costs and reduces performance. To remove this problem an adaptive and efficient dynamic model needs to be developed to readily characterize fairly precise positions incorporating interactive forces applied by actuators and external force elements. This will enable surfaces to be accurately simulated and the resulting model made available for model based control methods.

Kano et al. [10] developed the model of a two-dimensional sheet-like robot. Inspired by the decentralized control scheme of the scaffold-based locomotion of snakes, the surface undergoes large bends but it lacks information on the quality and quantity of induced curvatures and varying profiles. Medina et al. developed an actuated continuum surface using a shape memory alloy (SMA) material grid and further developed a model using Euler-Lagrange methods [11]. However, the array has poor scalability, as to improve performance and surface size the number of wires and nodes would be increased, resulting in dramatically increased computational demand. Also, SMA wires can withstand relatively small external forces, and are therefore, unable to be used for applications that require significant force interactions with the surface. Another approach is the use of the *phantom muscle method* by Merino et al. [12] to present a kinematic model for LDCSs using an infinite number of interpolated curves parallel to the single actuator when deformed by this actuator. This mathematical model is not computationally demanding, but it lacks to account for material properties, gravitational effects etc. The work presented here gives the development of a generalized LDCS model using a lumped-mass approach to model dynamic effects such as including gravitational loading, compliance, and the material properties etc. In addition, the model takes in-depth bending shear effects into account for thick plates undergoing large deformations through its specific two-layer configuration that has not been done or considered in previous studies by lumped-mass approaches.

In Sect. 7.2, the principles of formulation, assumptions, and relevant equations of the dynamical surface model using the lumped-mass approach is introduced. Section 7.3 details the experimental setup and validation technique under gravity. Section 7.4, compares and discusses the modelling and experimental results on the performance of the surface for two different boundary conditions. Finally, concluding remarks and future work to be carried out to enhance or promote the current development are provided in Sect. 7.5.

7.2 Lumped-Mass Approach to Model LDCS

This section explains the development of the surface model, which uses a lumped-mass approach. This modelling strategy has been extensively used for its consistency with large deformations, ease of implementation, avoidance of treating rigid and flexible motion in different ways, and its claim to represent in a generic way a range of systems of importance in engineering [13–15]. The work accomplished via this technique also exploits its considerable computational efficiency along with the reliable capture of system dynamics through low-order lumped models, e.g. [16, 17]. The modes of interest of, e.g., a flexible beam, can be modelled with a small number of undeformable masses, interconnected by springs and dampers [17]. For example, Kino et al. [18] and Torfs et al. [19] discretized a beam with a number of masses equal to the number of modes. The models predict very well the behaviour of the flexible beam and are well suited for controller design. However, the division of the beam into masses and the determination of the spring and damper constants is rather intuitive and empirical.

In this technique, lumped masses are linked together via linear springs and dampers. Each node/dot of the grid can be then subjected to its own weight, damping effects, and the reaction spring forces by the surrounding masses. Hence the momentum forces are transmitted to each mass due to the motion of neighbouring masses. Newton's second law and Hooke's law can be directly applied on the masses to work out the position of each mass in the next step rather than interpolating the points along the elements between two nodes and using shape functions to predict the deformed shape of the element which is carried out in FEM [20].

7.2.1 Surface Modelling and Assumptions

As shown in Fig. 7.2a, a three-dimensional, 2-layer lattice of lumped masses was developed in which the masses are interconnected together through linear springs and dampers. To clarify the connections of this 3D mass-spring-damper network, one corner of the surface has been magnified in Fig. 7.2b.

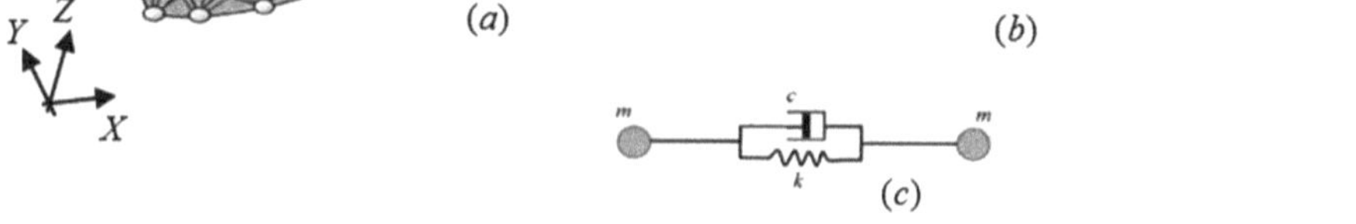

Fig. 7.2 Representation of the developed lumped-mass models; (**a**) 2-layer LDCS model, (**b**) close-up view of one corner of the surface model composed of 8 masses interconnected by liner springs, (**c**) The featured spring-damper link connecting every two masses in the model

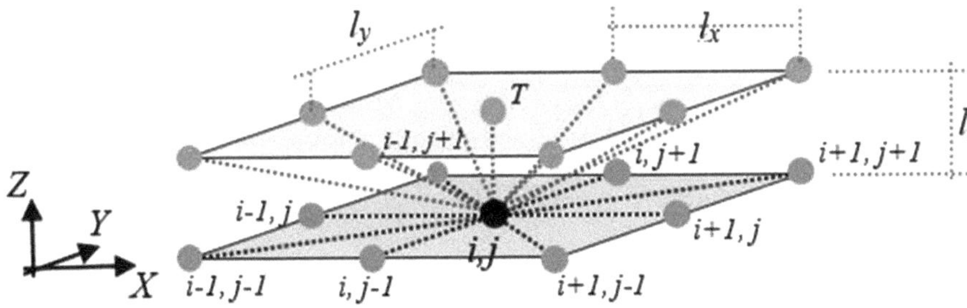

Fig. 7.3 Geometrical arrangement of the point masses connecting to the typical central mass *i,j* in the two-layer surface model

Because of the large surface deformations expected in LDCS, the planes of the surface that are perpendicular to the neutral plane at rest do not remain perpendicular after deformation. Therefore, the model includes diagonal springs in all available directions to enable shear effects to be considered and thus imply a more realistic behaviour of the actuated surface. As a result of this configuration a typical mass in the middle of one of the two layers is linked with eight surrounding masses in the same layer along with nine masses located on the opposite layer of the model. So a mass in the middle of one plane as shown in Fig. 7.3 is surrounded by 17 masses in total while the masses placed on surface edges and corners are linked to fewer of masses, i.e. 11 and 7 respectively. In this article any spring link is collocated with a damper of damping coefficient c as shown in Fig. 7.2c, however these are not depicted in the figures for clarity. The dampers are influential in providing the rate of energy absorption and suppression of the oscillations caused by loading and deformation of the surface.

The dynamics of a mass in the model is yielded by its own weight in combination with the forces exerted by surrounding springs. Moreover, some masses might be in contact with ambient systems that are then subject to external loading such as the interactive forces and moments applied directly by the actuators integrated with the surface. The model developed in this work contains few simplifying assumptions. The surface is not constrained in a particular axis. The velocity and acceleration of the masses are taken into account in all spatial Cartesian coordinates. As one of the main assumptions in this model, torsional springs have been neglected to be applied around the masses to avoid extremely added run time for solving the equations of motion related to angular momentum. This solution has been previously applied for modeling planar flexible mechanical systems such as 2D, beam-like arrays [15]. Instead, to provide shear effects, every mass is diagonally linked to the all surrounding masses which means a lumped mass in the middle of the surface (and not located on the edge) is interconnected with 12 masses only through diagonally positioned springs (see Fig. 7.3). As a result, the in-depth thickness can be applied to thicker surfaces. Also the surface model is flat and its main axes are straight before loading. Moreover, the material of the model in this work is assumed elastic, homogenous and isotropic.

7.2.2 Governing Equations

The overall dynamics of the 3D grid are given by the following equation which consists of matrix terms pertinent to all the existing masses

$$[M]\{\ddot{U}\} + [C]\{\dot{U}\} + [K]\{U\} = F(t) \tag{7.1}$$

where U stands for the vector of displacement, $[M]$ the inconsistent or diagonal lumped mass matrix, $[C]$ the damping matrix, $[K]$ the stiffness matrix, and $F(t)$ is the vector representing external forces (F_{ext}) and gravity (W) acting on the model i.e.

$$F(t) = F_{ext} + W \tag{7.2}$$

The spring force F_q caused by the transition of every two connected masses, according to Hooke's law, is determined as the following

$$F_q = k_q u_q^{12} + c_q \dot{u}_q^{12} \tag{7.3}$$

where q represents $\widehat{q} = \widehat{i}, \widehat{j}, \widehat{k}$ which are unit vectors for the Cartesian coordinates X, Y, and Z respectively. Likewise k_q and c_q are the stiffness of the spring and damping coefficient in direction q respectively, and u_q^{12} stands for the net stretch or compression of the spring along the coordinate q occurred after deformation which is determined through the following equation

$$u_q^{12} = \left[\left((l_x + u_{2x} - u_{1x})^2 + \left(l_y + u_{2y} - u_{1y}\right)^2 (l_z + u_{2z} - u_{1z})^2\right)^{1/2} - L\right] \times \left[\frac{\left(l_q + u_{2q} - u_{1q}\right)}{\left((l_x + u_{2x} - u_{1x})^2 + \left(l_y + u_{2y} - u_{1y}\right)^2 (l_z + u_{2z} - u_{1z})^2\right)}\right]^{1/2} \tag{7.4}$$

Moreover $\dot{u}_q^{12}$ in (7.3) is the time derivative of (7.4) defined as $\dot{u}_q^{12} = \frac{d}{dt}\left(u_q^{12}\right)$. The parameter u_{iq} ($i = 1$, 2 and $q =$ x, y, z) in (7.4) is the component of vector U in (7.1). Also, l_q stands for the distance between two masses in the direction q. Then a set of equations whose number depends on the number of lumped masses multiplied by 3 (the number of coordinates), can be solved numerically in parallel. Therefore, for the 2-layer grid of 9 × 9 lumped masses developed in this work, a mass matrix of 492 × 492 and a displacement vector of 492 components were built up accordingly to be solved numerically. The average simulation runtime for this set of equations was 225 s. All the equations of motion and numerical simulation were implemented using the commercial software MATLAB R2016a.

Figure 7.3 shows how a mass located in the center of one of the two surface's layer is bound to other masses by 17 linear springs.

In this illustration the surrounding masses are indexed by *i,j* abbreviations to identify positions and directions. Also, the index *T* stands for the masses located in the top layer. The equations of motion for the central point is then given by

$$
\begin{aligned}
u_q^{i,j} &= \int_{t_1}^{t_2}\left(\int_{t_1}^{t_2} \ddot{u}_q^{i,j} dt\right) dt = \frac{1}{m}\int_{t_1}^{t_2}\left(\int_{t_1}^{t_2} \sum F_q(t)dt\right) dt \\
&= \frac{1}{m}\int_{t_1}^{t_2}\left(\int_{t_1}^{t_2} \left\{k_{aq}\left[u_q^{i-1,j} + u_q^{i+1,j} + u_q^{i,j+1} + u_q^{i,j-1}\right]\right.\right. \\
&+ k_{dq}\left[u_q^{i+1,j+1} + u_q^{i+1,j-1} + u_q^{i-1,j+1} + u_q^{i-1,j-1}\right] \\
&+ k_{dq}^T\left[u_q^{(i-1,j)_T} + u_q^{(i+1,j)_T} + u_q^{(i,j+1)_T} + u_q^{(i,j-1)_T}\right] \\
&+ k_{ddq}^T\left[u_q^{(i+1,j+1)_T} + u_q^{(i+1,j-1)_T} + u_q^{(i-1,j+1)_T} + u_q^{(i-1,j-1)_T}\right] \\
&+ k_{zq}u_q^{(i,j)_T} + c_z\frac{d}{dt}\left[u_q^{i-1,j} + u_q^{i+1,j} + u_q^{i,j+1} + u_q^{i,j-1} + u_q^{i+1,j+1}\right. \\
&+ u_q^{i+1,j-1} + u_q^{i-1,j+1} + u_q^{i-1,j-1} + u_q^{(i-1,j)_T} + u_q^{(i+1,j)_T} \\
&+ u_q^{(i,j+1)_T} + u_q^{(i,j-1)_T} + u_q^{(i+1,j+1)_T} + u_q^{(i+1,j-1)_T} + u_q^{(i-1,j+1)_T} \\
&\left.\left.\left. + u_q^{(i-1,j-1)_T} + u_q^{(i,j)_T}\right] + F_{(\mathrm{ext})q} - W\right\} dt\right) dt
\end{aligned} \tag{7.5}
$$

where the terms u_q, (q = x, y, z) for the masses in all positions except the central mass (*i,j*) are determined using (7.4) noting that all the parameters l_x, l_y, l_z, and L can be given different values depending on the position of each mass relative to the mass *i,j*. In the current work, the assumption considered for the mass grid is $l_x = l_y = l$ while the thickness l_z can be quantified independently. Also, the spring stiffness varies in different directions depending on the positioning status of the springs. Accordingly, k_{aq} for $q = x, y, z$ in (7.5) stands for the stiffness of the *axial* springs in the planar layers in the direction q with the unloaded spring length of L= l and k_{dq} represents the stiffness of *diagonal* springs in plane with the unloaded length $L = \sqrt{2}l$. Likewise k_{dq}^T denotes the stiffness of the diagonal springs between the bottom and top layers with $L = \sqrt{l_z^2 + l^2}$; k_{ddq}^T the diagonal springs between the two layers but with the furthest distance of $L = \sqrt{l_z^2 + 2l^2}$; and finally k_{zq} is the stiffness of the spring connecting the mass (i,j) to $(i,j)_T$ with the unloaded length $L = l_z$. Any external forces in the model (e.g. those applied by actuators and gravity effects) are taken into account by $F_{(ext)q}$ and W respectively with the gravity force applied in the z-direction. Note that the terms corresponding to damping forces vanish in steady-state conditions and they affect the displacements only in transition states.

The equations for the masses located at the top layer are similar to (7.5) where the places of the top and bottom layers are swapped. Here the index T is replaced by B which indicates the masses located in the bottom layer of the model.

It should be noted that the equations of motion for boundary masses are different based on imposed conditions. When a mass is constrained in specific directions, the corresponding displacements must become zero rather than being calculated through (7.4). Likewise, for the masses linked to actuators, the amount of displacement is initially dictated by the motion of an actuator to which the mass is bound. In this case, some u_q terms in the far-right side of (7.5) is replaced by the displacements of corresponding points of the actuator.

7.3 Experimental Test Rig Setup

To validate the model for its application to LDCS, several statically deformed surface samples were constructed and their displacements were measured for comparison with the model. From these results, a suitable Young's modulus for the surface was determined through simple tensile tests. Then the experimental and lumped mass model results were compared for each test sample.

In the current paper two surface configurations are presented. Figure 7.4 depicts these two samples in the test setup; one with two clamped edges of zero curvature (Fig. 7.4a) and the other a sample with one clamped edge consisting of a steel rod which has been bent to the curvature 4.0 m^{-1} to represent the displacement of an actuator (Fig. 7.4b). Indicated in Fig. 7.4 are also the four points P_1, P_2, P_3, and P_4 whose displacements will be measured to be compared with modelling results. Point P_1 is located in the centre of the square surface, points P_2 and P_3 are on the middle of two sides of the surface and point P_4 is on its farthest corner.

Both flexible surfaces shown in Fig. 7.4 were cut in square shapes with the size of 0.188 m for each side. They were manufactured from a roll of nitrile rubber sheet of thickness 3 mm. The density of the rubber was found to be 1678 kg/m^3 by

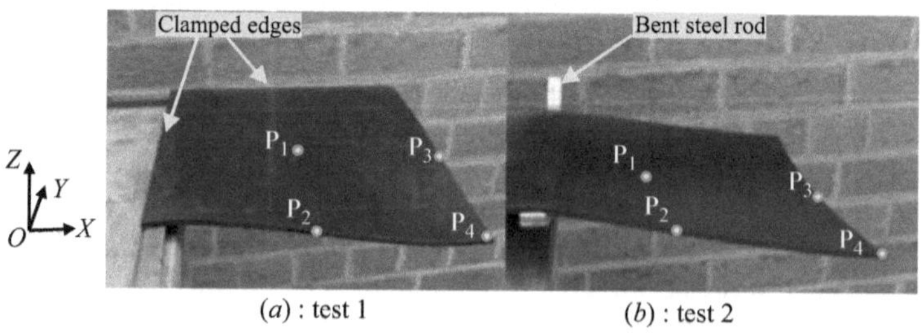

Fig. 7.4 Pictures of two test surface configurations; (**a**): Test 1- Surface of two perpendicular clamped edges with no curvature (**b**): Test 2- surface with one clamped edge bent by an actuator to the curvature 4.0 m^{-1}

measuring the mass and volume of several samples. For Poisson's ratio, a typical value of 0.48 for rubber was used. The tensile test to approximate Young's modulus was carried out on different sized rubber samples. This was done using an empirical approach based on maximum deflection of the rubber samples compared with the data produced by the surface model. A mean Young's modulus was eventually concluded as $E = 0.0545$ GPa.

To minimize the residual curvature of the rubber from storage on a roll, the surfaces were stacked and a 10 kg weight was placed on top for 48 h prior to assembly. Once bent the steel rod actuators were securely fixed to the rubber surface using double sided acrylic foam tape. To determine the displacement of desired points on the surface, *fringe projection techniques* were selected. A typical fringe projection system consists of a projection unit, a camera and a processing/analysis unit. The system used in the current study was composed of a laptop connected to a multimedia LCD projector, with a converging lens of 10 cm focal length, along with a PixeLINK PL-A741 CCD camera with a resolution of 1280 × 1024 pixels in grayscale. This method has been extensively investigated in the past and a summary of the technique with further details can be found in a review by Gorthi and Rastoghi [21]. The key advantage of this method is that it is non-contact and consequently the surface samples are not affected during measurement while high resolution 3D data for the points of interest are gathered almost instantaneously with measuring displacements up to 0.01 mm. A diagram of the setup used for our experimentis shown in Fig. 7.5.

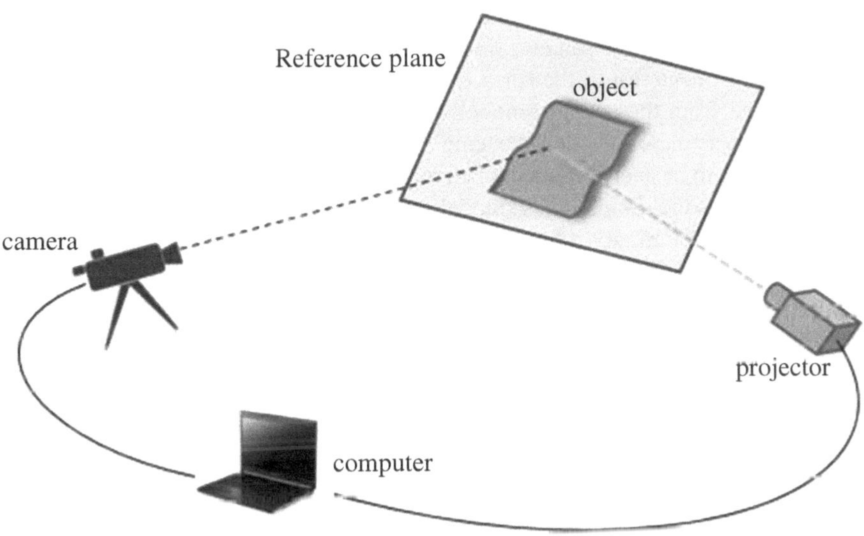

Fig. 7.5 Setup for fringe projection techniques

7.4 Comparison of Modelling and Experimental Results

7.4.1 Static Displacements

The first step in comparing the results was to initialise running the model within the solver with the material properties, surface parameters and boundary conditions for each test. Also the model coordinate system was adjusted according to the experimental test configurations. The surface model contains a 2-layer grid of 9×9 masses in each layer configured in the way described in Sect. 7.2. Therefore 162 masses of $m = 0.0011$ kg were utilised to build up the surface model and to conform to the experimental test surfaces. To approximate the spring constants of the LDCS model, an approach presented in [15], developed for lumped-mass systems, was utilised which worked out an average value of $k = 550$ N/m for the set of spring constants to match the experimentally determined Young's modulus.

For test 1 shown in Fig. 7.4a with two clamped straight edges, no actuator was needed for the surface model. For this reason, the masses on these two edges were simply constrained in all directions and the rest of the model was exposed to gravity force acting on each mass. In contrast, to implement test 2, an actuator model was required to deform the single clamped edge of the surface with the curvature 4.0 m^{-1} according to the curved surface in the test. Hence an actuator model was developed that matches the physical configuration of the surface model along that edge. The actuator was also modelled as a multi-body system with lumped segments of given mass and moment of inertia values linked by torsional springs and dampers developed through a Lagrangian formulation and set to approximate the desired curvature. The actuator was then integrated to the surface model to create any curvature of interest in the surface. In this way, the surface model is shown to be decoupled from the actuator model with only force interaction points required should other actuation models be employed. This also enables other actuation configurations to be easily applied when modelling LDCS.

For both tests the lumped-mass models were then loaded under their own weight to present the preliminary dynamic performance of the approach and to evaluate their consistency with the surface undergoing large deflections. Figure 7.6 presents the results of the modelled surfaces for test 1 and test 2. The displacements of the four points of the surfaces indicated in Figs. 7.4 and 7.6 have been compared in Tables 7.1 and 7.2 for test 1 and test 2 respectively. Table 7.3 presents a summary of the error between the experiment and simulation results for the given points.

General comparison between the model and experimental results presented in the tables reveals that the model successfully generated surfaces with comparable shapes for both tests 1 and 2. Carefully investigation of the modelling results identify the internal consistency of the model. For example, from the results of test 1 shown in Table 7.1, it is seen that as expected $| u_{x(\mathrm{P2})} | = - | u_{y(\mathrm{P3})}|$, $|u_{y(\mathrm{P2})} | = | | u_{x(\mathrm{P3})}|$ and $|u_{z(\mathrm{P2})} | = | u_{z(\mathrm{P3})}|$ which are intuitively rational based on the symmetry of boundary conditions and loading for the surface. Likewise, according to the results of test 2 shown in Table 7.2, the points P_1 and P_3 are not displaced at all in the y-

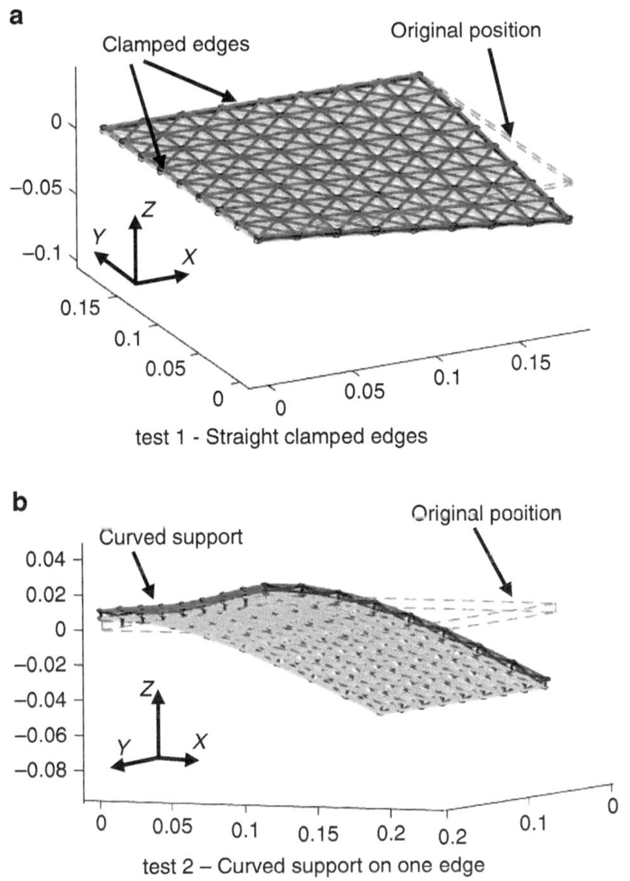

Fig. 7.6 Representation of lumped mass models under gravity for comparison with the experimental test surfaces shown in Fig. 7.4

Table 7.1 Comparison of the displacement results of the lumped mass model for a LDCS (Fig. 7.6a) with the experimental test 1 (Fig. 7.4a) at four different points of the surface. Data are in [mm]

Displacement (m)		P_1	P_2	P_3	P_4
Modelling	u_x	−0.33	−0.87	−0.76	−1.80
	u_y	0.33	0.76	0.87	1.80
	u_z	−6.02	−13.43	−13.43	−18.17
Experimental	u_x	−0.35	−0.81	−0.71	1.91
	u_y	0.34	−0.72	0.82	−1.86
	u_z	−6.45	−14.33	−14.33	−20.09

direction which is expected due to the symmetry about the surface's central x axis. As seen in Table 7.3, the point P_4 in test 1 undergoes the most notable difference in the Table but does not exceed 10% with mean percentage error of 4.8%. These errors are due to several factors including failure to remove all residual curvature from the rubber before assembly. This is roughly apparent in Fig. 7.4a where there is a minor

Table 7.2 Comparison of the displacement results of the lumped mass model for a LDCS (Fig. 7.6b) with the experimental test 2 (Fig. 7.4b) at four different points of the surface. Data are in [mm]

Displacement (m)		P_1	P_2	P_3	P_4
Modelling	u_x	−8.33	−8.01	−21.85	−20.48
	u_y	0.00	−0.05	0.00	−0.12
	u_z	−19.35	−14.29	−41.54	−31.34
Experimental	u_x	−7.94	−7.66	−20.75	−19.39
	u_y	0.00	−0.05	0.00	−0.11
	u_z	−18.87	−13.77	−38.88	−28.95

Table 7.3 Summary of error between modelling and experimental results at four points of the surface for both test 1 and test 2

% Error		P_1	P_2	P_3	P_4
Test 1	u_x	5.7	6.8	6.5	5.7
	u_y	2.9	5.2	5.7	3.2
	u_z	6.6	6.3	6.3	9.6
Test 2	u_x	4.6	4.3	5	5.3
	u_y	0	0	0	8.3
	u_z	2.4	3.6	6.4	7.6

unwanted twist in the experimental surface compared to the model. Another source of error is that because the surfaces were assembled by hand, the steel rods were not in the exact modelled position. This creates error in the surface shape which would be picked up in measurements, but not be present in the idealized model.

As with all other discretization techniques, the more nodes (here 'lumped masses') considered in the developed model, the more accurate the results achievable. But, on the other hand, the time consumed to construct the model as well as the computational burden to solve the problem will be increased. As a result, future work will need to consider the relative trade-offs between accuracy and computational cost. Overall, however these results indicate a reliable conformity despite the limited number of masses used in the model with a mean percentage error of 4.8% and relative resulting shapes remaining the same.

7.4.2 *Dynamic Analysis*

It should be noted that in the videos the surface model was run from zero gravity at the very beginning. For this reason, gravity switches abruptly from $g = 0$ to $g = 9.81$ m/s^2 after runtime starts, which causes the surface to undergo oscillation over time until it settles down at the final deflected position. This arrangement was programmed intentionally to simulate the dynamic behaviour of a soft surface when it is suddenly exposed to external forces such as gravity. However, this was not recorded in the experiments here and is the subject of future work. The magnitude of this oscillation can be managed by changing the material damping properties considered in the model. As an example of this however, Fig. 7.7 shows the displacement of point P_4 in test 1 in the three directions x, y and z.

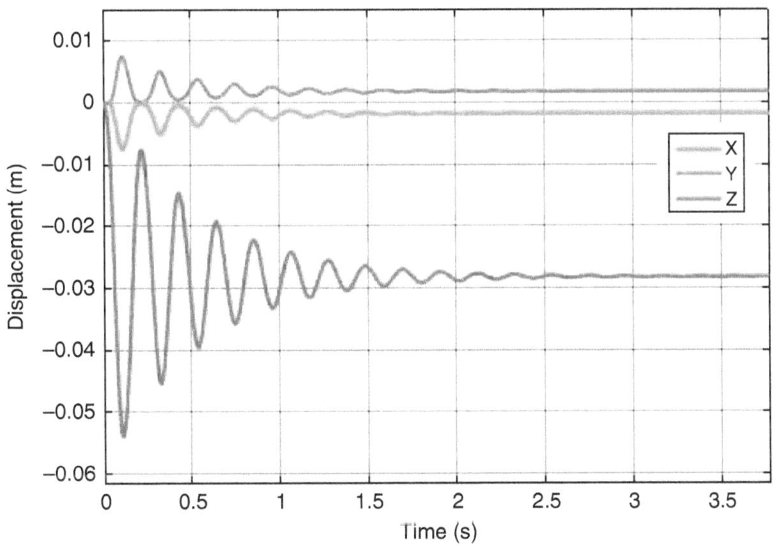

Fig. 7.7 Displacements of point P_4 on the surface in test 1 over time

7.5 Concluding Remarks and Future Work

In this work the application of a lumped mass approach to model 3D actuated Large Displacement Continuum Surfaces (LDCS) composed of mass-spring-damper arrays was presented. The developed model enables representation of behaviour of arbitrarily thick flexible plates to encompass in-depth and transverse shear effects. The model is also adaptable for integration with a variety of actuation elements to bend the surface into various curvatures.

A test rig was set up and two experimental surface configurations were tested as part of initial validation efforts. The configurations showed satisfactory agreement between the experimental and model results for such a flexible surface undergoing uniformly distributed gravitational loads. The mean percentage error was found to be 4.8%, while the results matched the expected performance.

Given the promising results obtained here for both LDCS configurations, further research to improve the model is motivated and ongoing. For example, new materials to form the LDCS are being evaluated and tested to see if one could overcome the material issues encountered here while matching the assumptions made in the current model. Additionally, ongoing work will extend the 3D lumped surface model integrated with the actuator models to deliver a robust dynamic model for actuated LDCS that will precisely describe the surface profile for both displacement of any point across the surface, as well as interactive forces applied by actuators and external forces in a range of configurations. Further, experiments will be carried out on multiple actuator configurations, where the motion of the flexible surface is constrained by additional, more complex boundary conditions. This will enable into a longer term research looking to develop accurate and efficient control of actuated continuum surfaces undergoing large deformations.

Acknowledgement This work is funded and supported by the Engineering and Physical Sciences Research Council (EPSRC) under grant number: EP/N022505/1, and the Natural Science Foundation of China (NSFC) under grant number: 51611130202.

References

1. Burgner-Kahrs, J., Rucker, D.C., Choset, H.: Continuum robots for medical applications: a survey. IEEE Trans. Robot. **31**(6), 1261–1280 (2015)
2. Santiago, J.L.C., Walker, I.D., Godage, I.S.: Continuum robots for space applications based on layer-jamming scales with stiffening capability. In *IEEE Aerospace Conference*, pp. 1–13 (2015)
3. Bajo A., Simaan, N.: Finding lost wrenches: using continuum robots for contact detection and estimation of contact location. In IEEE International Conference on Robotics and Automation, pp. 3666–3673, Anchorage, Alaska (2010)
4. Walker, I.D.: Continuous backbone "continuum" robot manipulators. ISRN Robot. **2013**, 1–19 (2013)
5. Godage, I.S., Medrano-Cerda, G.A., Branson, D.T., Guglielmino, E., Caldwell, D.G.: Dynamics for variable length multisection continuum arms. Int. J. Robot. Res. **35**(6), 695–772 (2016)
6. Conrad, B.L., Jung, J., Penning, R.S.: Interleaved continuum-rigid manipulation: an augmented approach for robotic minimally-invasive flexible catheter-based procedures. In: IEEE International Conference on Robotics and Automation, pp. 718–724 (2013)
7. Webster, R.J., Jones, B.A.: Design and kinematic modeling of constant curvature continuum robots: a review. Int. J. Robot. Res. **29**(13), 1661–1683 (2010)
8. Tatlicioglu, E., Walker I.D., Dawson, D.M.: Dynamic modelling for planar extensible continuum robot manipulators. In: IEEE International Conference on Robotics and Automation, pp. 1357–1362, Roma (2007)
9. Bailly, Y., Amirat, Y., Fried, G.: Modeling and control of a continuum style microrobot for endovascular surgery. IEEE Trans. Robot. **27**(5), 1024–1030 (2011)
10. Kano, T., Watanabe, Y., Ishiguro, A.: Sheetbot, two-dimensional sheet-like robot as a tool for constructing universal decentralized control systems. In: IEEE International Conference on Robotics and Automation (ICRA), pp. 3733–3738 (2012)
11. Medina, O., Shapiro, A., Shvalb, N.: Minimal actuation for a flat actuated flexible manifold. IEEE Trans. Robot. **32**(3), 698–706 (2016)
12. Merino, J., Threatt, A.L., Walker I.D., Green, K.E.: Forward kinematic model for continuum robotic surfaces. In: IEEE/RSJ International Conference on Intelligent Robots and Systems (IROS), pp. 3453–3460, Vilamoura, Algarve, Portugal (2012)
13. Williams Jr., J.H.: Fundamental of Applied Dynamics. Wiley, USA (1996)
14. Mesit, J., Guha, R., Chaudhry, S.: 3D soft body simulation using mass-spring system with internal pressure force and simplified implicit integration. J. Comput. **2**(8), 34–43 (2007)
15. Habibi, H., O'Connor, W.: Wave-based control of planar motion of beam-like mass–spring arrays. Wave Motion. **72**, 317–330 (2017)
16. Liu, T., Bargteil, A.W., O'Brien, J.F., Kavan, L.: Fast simulation of mass-spring systems. ACM Trans. Graph. **32**(6), 1–7 (2013)
17. Anthonis, J., Ramon, H.: Comparison between the discrete and finite element methods for modelling an agricultural spray boom – part 1: theoretical derivation. J. Sound Vib. **266**, 515–534 (2003)

18. Kino, M., Murakami, T., Ohnishi, K.: A vibration suppression and disturbance rejection strategy of flexible manipulator by multiple acceleration feedback. In: Proceedings of Movic '98, Vol. 2, pp. 781–786, Zurich (1998)
19. Torfs, D.E., Vuerinckx, R., Swevers, J., Schoukens, J.: Comparison of two feedforward design methods aiming at accurate trajectory tracking of the endpoint of a flexible robot arm. IEEE Trans. Control Syst. Technol. **6**(1), 2–14 (1998)
20. Petyt, M.: Introduction to Finite Element Vibration Analysis. Cambridge University Press, Cambridge (1990)
21. Gorthi, S.S., Rastogi, P.: Fringe projection techniques: whither we are? Opt. Lasers Eng. **48**(2), 133–140 (2010)

Chapter 8
Research and Development of Methods and Tools for Rapid Digital Simulation and Design of Personalized Orthoses

Iliya Savov, Georgi Todorov, Yavor Sofronov, and Konstantin Kamberov

Abstract The presented work covers the definition of three different methods for designing orthoses produced by the methods of rapid prototyping of polymer structures. The stages of design and production of orthoses are defined. The methods are illustrated, by way of example, of an orthosis for upper limb fracture. The use of parametric modeling facilitates easier and faster preparation of CAD models. In the production of orthoses, the use of an automated approach by adapting a parametric model leads to a number of advantages. Besides shortening the required workflow time, there are opportunities to track the effectiveness and performance of the end products and adjust the relevant parametric models in the catalog in order to improve them and embed the resulting know-how in the model. The creation of a workflow to facilitate easier CAD modeling in combination with reverse engineering technology and rapid prototyping technologies offers excellent opportunities to improve the resulting orthoses for the patients in need.

Keywords Reverse engineering · Rapid prototyping · Rapid design · Orthoses

8.1 Orthoses Overview

Healthcare is an area of increasing importance to the quality of life. One of the serious problems in this regard is the ability to achieve efficiency gains and quality of implants, prostheses and orthoses for patients in need of these items.

I. Savov (✉) · G. Todorov · Y. Sofronov · K. Kamberov
Faculty of Industrial Technology, Technical University Sofia, Sofia, Bulgaria
e-mail: gdt@tu-sofia.bg; ysofronov@tu-sofia.bg; kkamberov@3clab.com

E. Zahariev, J. Cuadrado (eds.), *IUTAM Symposium on Intelligent Multibody Systems – Dynamics, Control, Simulation*, IUTAM Bookseries 33,
https://doi.org/10.1007/978-3-030-00527-6_8

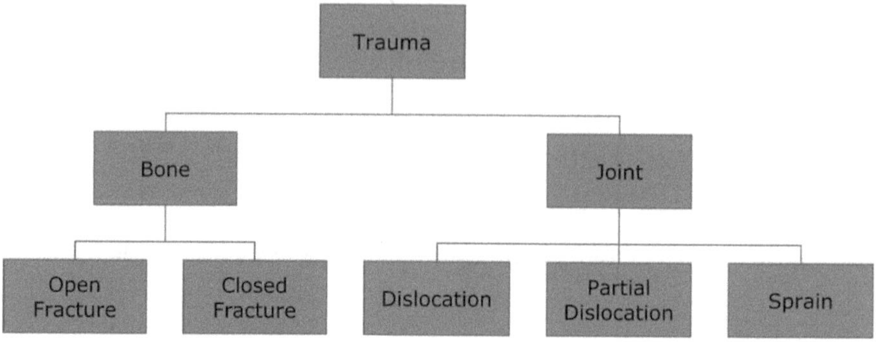

Fig. 8.1 Classification of trauma

8.1.1 Classification of Trauma

There are several major types of injuries that require the use of orthoses or implants as part of the patient's treatment. They could be classified into two large groups – bone and joint. Bone injuries are two main types of fractures [1]:

- Open fractures – loss of continuity of bone tissue accompanied by an open wound.
- Closed fractures – loss of continuity of bone tissue where the skin is not injured or the wounds are superficial and unrelated to the fracture.

Joint injuries are:

- Luxations – disturbance of the location and contact of the bones forming the joint with each other.
- Subluxations – the location of the contact surfaces of the bones forming the joint is impaired, but the loss of contact is incomplete
- Sprains – partial rupture of tendons responsible for joint stability (Fig. 8.1).

8.1.2 Treatment of Trauma

As part of the treatment of all these traumas, it is necessary to stabilize and partially or completely immobilize affected bones or joints. Methods to achieve this can be defined as operative – by surgical nails and other implants, and non-operative – by different orthoses. The rapid development of medical and industrial scanning equipment, computing, CAD/CAM software packages, direct digital production methods, and new materials allow for a significant improvement in these medical devices, which have not undergone any significant changes of the principles of operation since the dawn of medicine. The focus of this research is on the development of the most commonly used elements for non-operative limb stabilization – orthoses.

Orthopedic splints for the upper limbs are those which immobilize the hand and/or wrist. In the construction three contact points are created, creating resistance in the direction opposite to the displacement of the bone. Orthopedic splints of the lower limbs are those which immobilize the foot and/or leg. In all types of orthopedic splints examined, the same principle of acting to apply fracture stabilizing force through the soft tissues surrounding the injured bone is present. Different types of splints, regardless of their place of application and their type, generally have the same requirements for the manufacturing technology.

8.1.3 Manufacturing Technology

Technologies to prepare orthopedic splints are determined largely by the requirements of this type of product. They should be made of biocompatible materials, provide the necessary force to stabilize the fracture and be consistent with the relevant physiological and anthropological parameters of the patient.

Historically, the most commonly used method of production for these articles is building a gypsum shell by wrapping the limb with layers of gypsum textile strips. The process begins with the application of several protective textile layers to protect the patient's skin from injury and to reduce the sense of discomfort. Then gypsum lint and/or bandages are placed and wrapped with well moistened bandages. After drying the thus formed shell, an individual splint is obtained which corresponds to the physiological parameters of the patient and is adapted to the particular limb. Among the advantages of this method is the low cost of gypsum orthotics and the lack of need for special equipment for its preparation. The main drawback of this approach is the lack of access to the limb during treatment that does not allow the removal of dead tissue on the surface of the skin. This can lead to itching, infections, exhaustion, burns, allergic reactions and other complications. Another disadvantage is the high weight of the gypsum shell, as well as the large volume that makes the patient's everyday life difficult. Removal of the splint requires its destruction, which is a noisy process that creates discomfort in the patient and poses the risk of injury. This is still the most common method of treatment.

Standard gypsum shells have been improved by using more advanced materials. The use of polymeric resins results in a reduction in the weight and volume of the produced shell. They are made from resin-impregnated cotton, fiberglass or polypropylene. This also leads to good water resistance of the final product. However, the disadvantages associated with the lack of access to the skin in the affected area and the manner of removing the splint remain unresolved.

In certain cases, it is possible to use various pre-made standard orthoses in the later stages of treatment and sometimes throughout the whole process. They usually consist of shells made of polymeric materials or metal bars that are attached to the limb by means of textile bandages and fasteners. These devices allow for the orthosis to be removed and placed back on the limb and usually have a lower weight than the gypsum shells. Their main disadvantage is that they are made in a standardized form and size and often do not meet the individual needs of the patient. They can be adapted to a certain extent by deforming the standard product, but the possibilities are limited. [1]

8.1.3.1 Classification

Classical manufacturing technologies for orthoses can be classified in two large categories:

- Handmade – Those technologies rely on the manual labor of a highly qualified and experienced technician who produces the orthosis by hand. Their main advantage is that they are personalized to the patient who needs them, and their main disadvantages are related to the limitations determined by the materials used and the production techniques which require the orthosis to be directly modelled onto the patient's limb.
- Industrial – Those technologies use much wider array of materials with superior properties and allow advanced features such as removal and re-placement of the orthosis, complete water resistance, continuous adjustment of the applied force, better airflow to the patient's skin, reduced weight and volume. All of these advantages, however come at the cost of them being standardized devices which are not produced in accordance with the specific patient's limb geometry and may not fit comfortably or at all despite their advantages over standard casts.

The disadvantages of both types of classical manufacturing technologies suggest the need for further improvement of the process of production for orthotic devices. This has led to the emergence of new modern manufacturing processes based on rapid prototyping technologies. They are personalized, designed for the specific patient, thus having the same advantages of the classically produced personalized orthoses, while in the same time allowing the use of more advanced materials, such as thermoplastic polymers. Thus, they combine the advantages of both of the classical manufacturing methods. Because the orthosis produced in such a manner is personalized for the specific patient and does not require the same adjustment features that standardized orthoses need, the volume and weight of the device can be reduced even further [2–4] (Fig. 8.2).

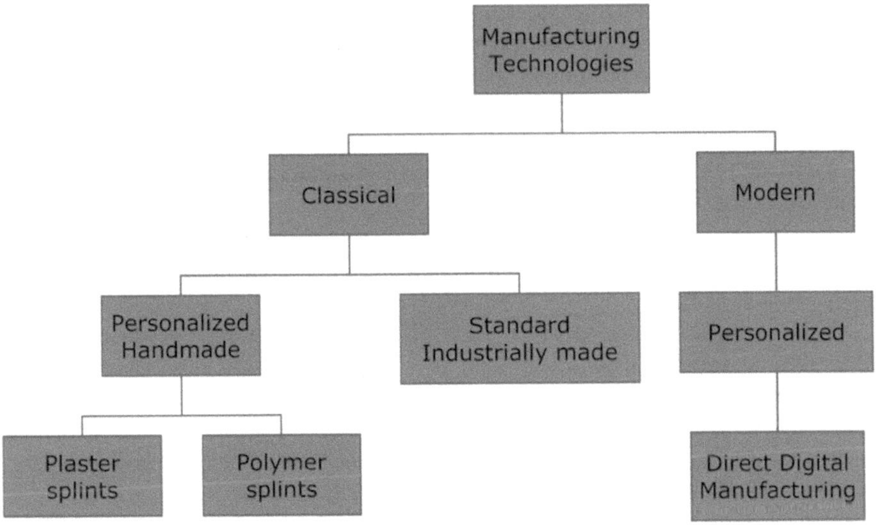

Fig. 8.2 Classification of manufacturing technologies

8.2 CAD Design of Orthoses

8.2.1 Reverse Engineering in Medicine

The first step in the process of reverse engineering is determining the required input data. The choice determines not only how data is collected but also determines the methods of processing and analysis used. The choice of input data has a significant effect on the level of accuracy of the constructed three-dimensional model and, as a result, on the final product. Depending on the final application, different inputs, such as biomedical objects and samples, molds and castings, are used as well as patients themselves. From them the geometric information about the investigated objects is collected for the purpose of development of medical devices and research. Input data should be selected to meet both technical requirements and clinical limitations [5].

8.2.1.1 Scanning

There are different methods for collecting data for the needs of medical reverse engineering. They can be classified into two main groups.

- Contact
- Contactless

Contact methods use mechanical hands, coordinate measuring machines and CNC surface digitizers. Non-contact methods are based on two-dimensional images and clouds of points representing the geometry of the object that are captured by generating energy (light, sound, magnetic fields), and monitoring the transmitted or reflected result. There is no contact between the object under investigation and the device used in the reverse engineering process. The end result of the process is data from one of two types:

- Cloud of points
- Two-dimensional cuts

When using laser scanners and structured light scanners, the result is a cloud of points, and using techniques such as computed tomography leads to two-dimensional cross-sections [2, 6].

In orthosis creation, the input data is the geometric data for the patient's limb needed for treatment. Due to the need for them to be taken directly from the patient, contact methods for scanning are not suitable for administration. It is possible to use a laser scanner or a structured light scanner that results in a cloud of dots. Another option is to apply a computer tomography (CT) or magnetic resonance imaging (MRI), which presents sections of the examined limb in depth and provides information on the structure and condition of the tissues and bones under the surface. In this sense, CT and MRI lead to a larger and more complete set of data [7]. Those types of scanners are available in medical facilities and medical practitioners have experience with using them and processing the scanning results. However, the data processing methods as well as the technological and medical requirements for the orthoses developed suggest that surface methods offer sufficient information for the modeling of the orthosis but have the advantage of simple data capture and processing.

8.2.1.2 Processing of Scanned Data

Depending on the data collection method used, it is necessary to apply different approaches to processing until the final model is obtained. The process largely depends on the particular scanning device being used and the accompanying software offered by the manufacturer [5, 8].

Broadly speaking, when using a traditional medical scanning technology, the different scanned images are segmented and the resulting volumes are exported as meshes. When using surface scanning technologies, the resulting cloud of points is filtered and converted to a mesh as well. Regardless of the technology used, the end result of the data processing is geometric representation of the object in the form of a network of triangles or if using a more advanced post-processing tool a NURBS CAD model. During the process of designing of the orthosis further post-processing of the scanned data may be needed to facilitate the needs of the CAD software [7].

8.2.2 *Design of Orthoses*

Once the geometric model of the examined limb has been obtained, the model of orthosis should be constructed. There are several possible modeling approaches, and the specific choice depends on the capabilities of the CAD model used.

In the recent years direct modelling solutions have become increasingly popular among both designers and medical practitioners. They offer some advantages in terms of ease of use, especially when creating complex geometric shapes in comparison with traditional parametric CAD modelling tools. However, the focus of this work are approaches based around parametric modelling. The reason for this choice is because, while providing some advantages, direct modelling approaches provide less options for automation of the processes in comparison to parametric approaches. While the process of modelling an individual orthosis has some challenges yet to be conquered, the focus of this study is in developing methods for rapid, highly automated design, that would provide the mass adoption of reverse engineering and rapid prototyping technologies in the production of orthoses [3, 6, 9, 10].

8.2.2.1 Sections

One of the possible approaches for designing orthoses is by designing cross-sections and connecting them to obtain the final product (see Fig. 8.3). Many of the available

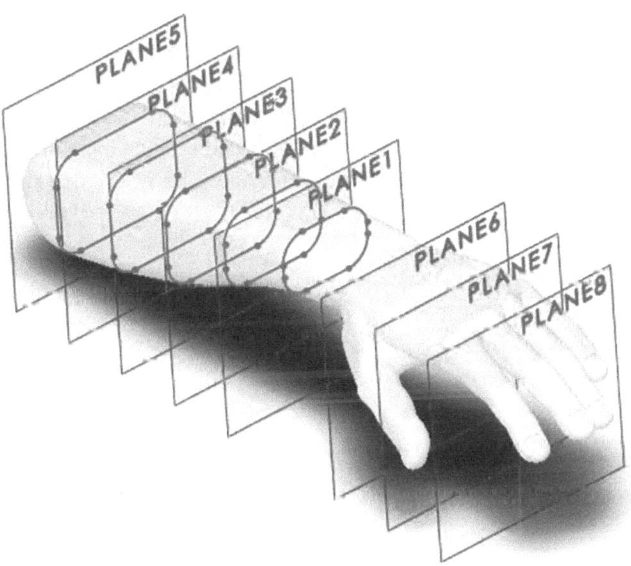

Fig. 8.3 Design of orthoses via sections

CAD modelling softwares offer the ability to automatically generate sketches of the section of the object, which would help to facilitate the construction of cross-section sketches corresponding precisely to the geometric features of the treated limb – based on the scanned model data. This approach offers good control to the designer over the shape of the final product, while providing the necessary link with the geometric data collected by the object. At the same time, the characteristics of the shape of the human limbs, namely smoothness, continuity and smoothness, mean that the number of sections required to achieve the desired precision of the modeling orthotics is comparatively low [11]. The number of cross-sections needed can be further reduced by using different cross-section density in different areas, depending on the accuracy needed for each of them. This design approach mirrors the principles used in medical imaging technology, namely the use of two-dimensional cross-sections to represent a three dimensional object. An approach that has proven to provide results of sufficient accuracy for medical purposes. After the basic form is created, the orthosis is to be divided into parts in order to allow insertion and removal of the patient's limb and to add the corresponding attachments that fasten the individual parts [12]. At the end of the orthosis modeling, the limb model must be removed and the final device model is thus obtained.

8.2.2.2 Surface offset

Another possible approach, that requires a higher level modeling software with good surface modeling capabilities is to use the surface of the scanned limb as the basis of the orthosis. By offsetting the surface at different distances it can be used to define the inner and outer boundaries of the volume of the orthosis (see Fig. 8.4) An advantage of this approach is that the resulting form of the device derives completely from the geometric data obtained from the scan. However, this method brings with it some difficulties. The possibilities for control of the shape of the article are relatively limited – it is not always possible to edit the displaced surface in a precise way [11]. After defining the basic shape of the article, it should be re-divided into parts and fasteners should be added.

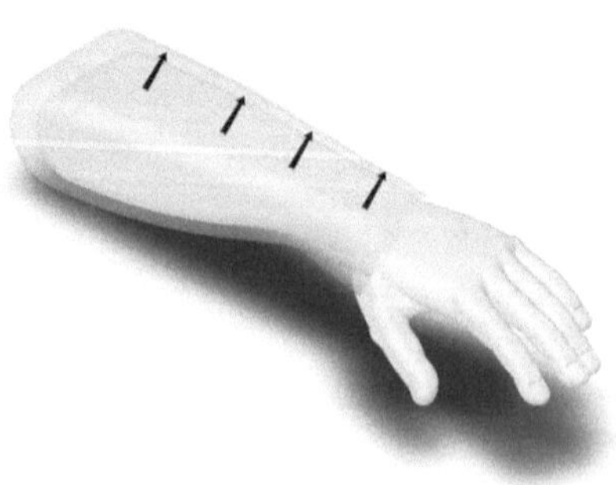

Fig. 8.4 Design of orthoses via offset of surfaces

8.2.2.3 Premade Models

Another possible approach is the use of premade models of the orthosis containing all of the required features built into them. The model is essentially a stock for the orthosis from which the final device is created. For that purpose, a family of models with different sizes is designed. The size of the models can be adjusted to a certain degree by using a few parameters. The scanned and reconstructed geometry is then cut from the model thus producing the required geometry of the internal surface of the model in order to fit the specific patient. No further design of features such as fasteners is needed as they are already built into the model.

This approach has been tested in detail in different parametric CAD packages and has proven to be feasible in any of them. It is possible to construct the premade model using either surface modelling or solid modelling. The choice of modelling technique is largely dependent on the design goals of the particular premade model and the personal preferences of the designer. When designing the premade object, the data of one or more patient scans can be used as a reference, but it is best to keep the model separate from any specific patient data and make the model independent from it. By designing the premade model to be completely detached from the scanned data the model is more robust and the chance of failure of the model is decreased.

Using that approach offers some significant advantages. When designing an orthosis from scratch the result is highly dependent on the experience of the technician who's designing it. This is very similar to the classical gypsum cast creation method in its approach to the design and production of the device. In both cases the results of the process are dependent on the subjective skills and experience of the person applying the approach. In comparison when utilizing premade models each model can be revised and improved to integrate any knowhow acquired and feedback from real world testing of the product (Fig. 8.5).

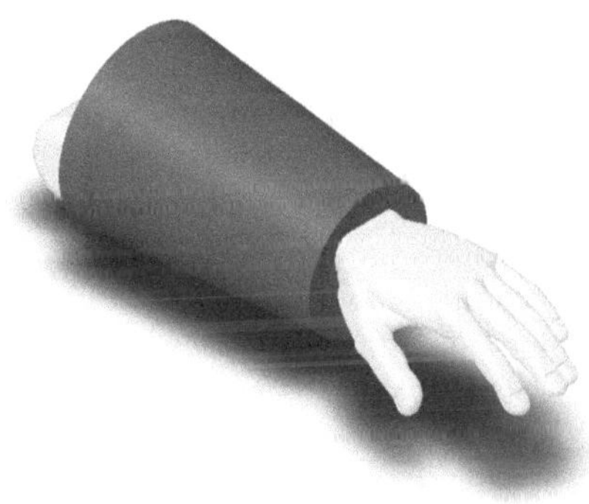

Fig. 8.5 Design of orthoses via premade models

8.3 Rapid Design

One of the obstacles to the wider application of reverse engineering methods and the achievements of modern technologies for rapid prototyping in medicine is the need for a large amount of multidisciplinary knowledge to be successful. One way to deal with this problem is by automating parts of the processes in the workflow [13].

Inherently, bone fracture orthoses are devices that have to match the patient's individual features but are simple and similar in nature. Their purpose is to stabilize the treated bone to facilitate its healing. Since the limbs of all patients have practically identical topology and a very similar shape, and the dimensions vary within predictable limits, it is possible to standardize and automate the process of orthosis creation. In the process of CAD modeling parametric models are commonly used which allows to have families of products easily generated just by changing the parameters of the model. For automating modeling of orthosis, it is necessary to define the number of parameters that serve to control the size of the orthosis model to adapt it to the data obtained through reverse engineering the patient's limb. Generally, the number of parameters depends on the complexity of the form and the amount of control needed on the parametric model.

When using a design approach via sections, the number of parameters depends on the number of sections and the control over form. Two approaches are possible for the control of the separate sections:

- By displacement of the loop – the contour of the drawing is shifted a certain distance from the nominal designed contour. In this approach, the effect of scaling individual sections by the relevant factors makes possible amendment only in size but not in the form of the section.
- By control points – defined as the respective number of key points through which the curve defining the shape of the particular section is controlled. By changing the parameters determining the location of points, both the size and shape of the contour can be controlled.

Both approaches have their advantages and disadvantages. Although it offers a higher level of control over the properties of the product being modelled, the approach relying on control points requires the use of a significantly larger number of parameters. With the increase of the parameters used it becomes increasingly more difficult and complicated to adapt the model to the scanned data. Using offset contour implies significantly less parameters, but in certain situations does not offer the required level of control. To achieve the optimal balance a hybrid approach can be used – control points in areas that require more precise control over the shape and contour shift in areas where it is not necessary.

If a modeling approach by offset of the surface is used it will not achieve the necessary control over form – only the distance of displacement can be managed. In this approach it is necessary to substitute the scanned data used as a basis for the construction of the model with the new scan and the whole model to be regenerated.

This approach would offer better accuracy of the results, but requires modeling software of the highest level and in case of problems and disintegration of the CAD model options for correction are very limited.

If a modeling approach using premade models is used the amount of parameters needed to adapt the model is limited. Those parameters are mostly related to controlling the overall dimensions of the orthosis. Using some secondary parameters, the outer shape of the model can be controlled to an extent if necessary. When the model is sized as needed the next step is forming the cavity inside it with the proper form. This is achieved by aligning the reconstructed scan results with the premade model and cutting them out of it. This approach offers a very simplified process of creation of the orthosis and guarantees that it will fit the patient perfectly. It essentially divides the device in two parts, based on its surfaces. The inner surface is the functional surface of the orthosis. The approach puts emphasis on the precision of the shape and dimensions of that surface and derives it entirely from the scanned results. Thus as long as the scanned data has the needed accuracy the resulting device will have it as well. The outer surface of the device is not in contact with the affected area and does not directly participate in the treatment of the patient. It rather has a supportive role and determines the overall dimensions of the orthosis. The requirements for it are mainly related to providing sufficient thickness of the shell while maintaining the volume and weight of the device as low as possible as to reduce the discomfort for the patient as much as possible.

This approach provides the necessary accuracy only where it's really required by emphasizing on the simplicity of the workflow. Design skills and knowhow are invested in the premade model instead of into the final orthosis, thus increasing the efficiency of the design process and iteratively perfecting the design. If there is need for further customization it can still be done, but a better approach would be to create a new premade model to fit the specific requirements. By storing all of the different premade models in a catalogue the types of patient cases that the collection of orthoses models covers increases. An example of this approach is shown on Fig. 8.7. The premade model used in this example has been designed for a patient's right forearm. It can be used for a left forearm as well via mirroring of the premade model. The initial model can be seen on Fig. 8.6 The model in the example uses six main parameters to size it properly and has ten additional parameters in cases where more precise control of the shape is needed. A scanned forearm and hand

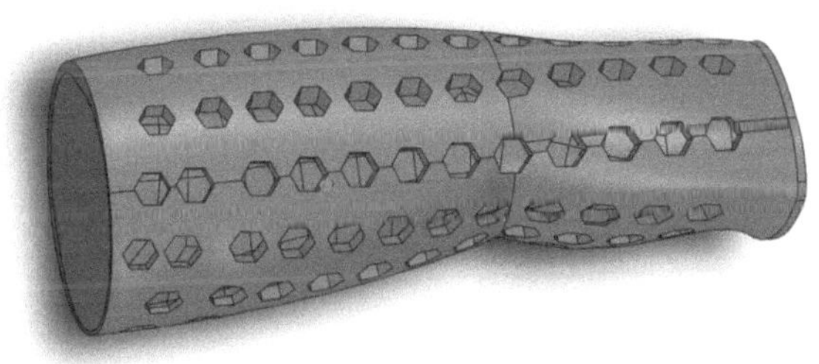

Fig. 8.6 Example of a premade model for orthosis creation

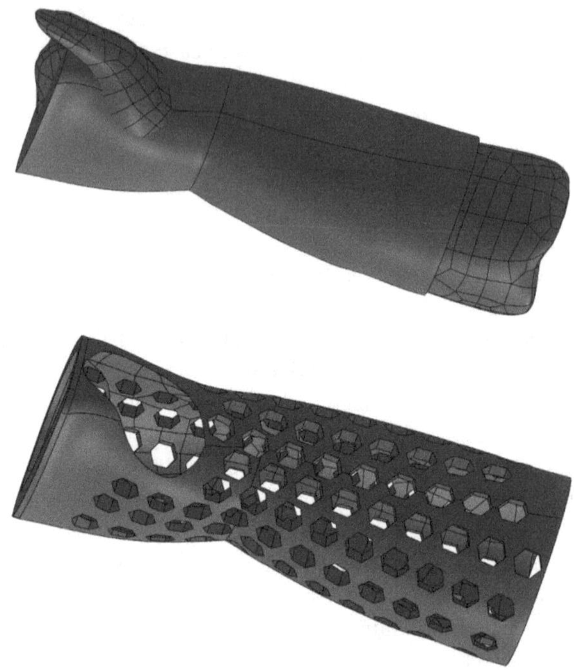

Fig. 8.7 Example of orthosis creation using premade models

is placed inside it (as seen on Fig. 8.7, where the perforation of the orthosis has been disabled to better show the intersection of the scanned data and the model). After cutting the scanned data out of the premade model the orthosis is finished. There is some excess material at different parts of the model, that would not be present if the orthosis was designed from scratch, however the results satisfy all design and functional requirements while simplifying and speeding up the design process as much as possible. The resulting device is lightweight and perforated to allow ventilation.

The same approach allows for the creation of more complex orthoses models with further improvements to the design, such as better perforation patterns and more complex basic shapes. For best results and fastest orthoses creation time it is best to always scan the patient limbs using the same orientation to facilitate easier alignment with the premade model. The alignment step of the process can be further improved by using different algorithms for automatic or assisted alignment.

The approach using premade models facilitates the rapid design of orthoses and should increase the adoption rate of the technology in real world applications, transitioning it from a niche solution and reducing the multidisciplinary knowledge required to implement it.

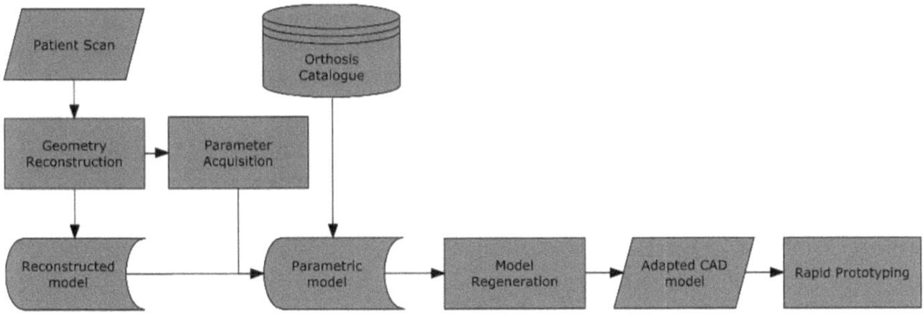

Fig. 8.8 Scheme of the rapid orthosis design process

8.3.1 Workflow

The principle scheme of the automated orthosis creation workflow is illustrated on Fig. 8.8. The process begins by scanning a patient's injured limb. This step produces the input data for the modelling, which is processed via geometry reconstruction. The reconstructed geometry itself provides the reconstructed model, which is used in the further steps of the creation process. Any required parameters for the configuration of the used premade parametric model are derived from the reconstructed model. Using an orthosis catalogue a suitable premade model is selected. The parameters derived from the reconstructed geometry model are used to adjust the parametric model in order to fit the patient. The reconstructed model and the premade model are combined together, by cutting the scanned data out of the orthosis from the catalogue. Thus, the internal contact surface of the device is derived from the patient geometry while the rest of the orthosis is designed in advance with all other functional requirements taken into account. The model is regenerated with the new data and this leads to the output of the process, which is the adapted to the specific patient CAD model. The resulting model can be produced by any available rapid prototyping model technology that provides the necessary strength of the prototype and has sufficient working area to produce the model. Any new knowhow and feedback acquired during the use of the produced model can be implemented in a new version of the parametric model in the orthosis catalogue.

8.4 Conclusion

Healthcare is an area of increasing importance to the quality of life. One of the serious problems in this regard is the ability to achieve efficiency gains and quality of implants, prostheses and orthoses for patients in need of these items.

The presented work covers the definition of three different methods for designing orthoses produced by the methods of rapid prototyping of polymer structures. The stages of design and production of orthoses are defined. The methods are illustrated, by way of example, of an orthosis for upper limb fracture. The use of parametric modeling facilitates easier and faster preparation of CAD models. In the production of orthoses, the use of an automated approach by adapting a parametric model leads to a number of advantages. Besides shortening the required workflow time, there are opportunities to track the effectiveness and performance of the end products and adjust the relevant parametric models in the catalog in order to improve them and embed the resulting knowhow in the model. The creation of a workflow to facilitate easier CAD modeling in combination with reverse engineering technology and rapid prototyping technologies offers excellent opportunities to improve the resulting orthoses for the patients in need.

The paper presents three different approaches for the construction of orthoses and a workflow for the process of mass adoption of rapid design techniques. While the paper is focused on one of the approaches it is possible to create designs which combine elements of all of them. The proposed workflow is flexible enough to facilitate any of the design methods and to be expanded further. The workflow and design approaches are also applicable more broadly in the design of implants and prostheses as well, since all those types of medical devices share some similarities in terms of their functional requirements.

Acknowledgments This research study is performed by the support of project BG05M2OP001-1.001-0008 “National Center for Mechatronics and Clean Technologies” of “Science and Education for Smart Growth” Operational Programme Executive Agency, Ministry of Education, Youth and Science, Bulgaria.

References

1. McRae, R., Esser, M.: Practical Fracture Treatment. Churchill Livingstone, Edinburgh (2008)
2. Gerber, N., Stieglitz, L., Peterhans, M., et al.: Using rapid prototyping molds to create patient specific polymethylmethacrylate implants in cranioplasty. Conf Proc Annu Int Conf IEEE Eng Med Biol Soc IEEE Eng Med Biol Soc Annu Conf 2010, pp. 3357–3360 (2010). https://doi.org/10.1109/IEMBS.2010.5627903
3. Dombroski, C.E., Balsdon, M.E., Froats, A.: The use of a low cost 3D scanning and printing tool in the manufacture of custom-made foot orthoses: a preliminary study. BMC. Res. Notes. **7**, 443–443 (2014). https://doi.org/10.1186/1756-0500-7-443
4. Payne, C.: Cost benefit comparison of plaster casts and optical scans of the foot for the manufacture of foot orthoses. Aust. J. Podiatr. Med. **41**, 29–31 (2007)
5. Hieu, L.C., Sloten, J.V., Hung, L.T., et al.: Medical reverse engineering applications and methods. 2ND Int Conf Innov Recent Trends Chall Mechatron Mech Eng New High-Tech Prod Dev MECAHITECH 10, pp. 232–246 (2010)
6. Toshev, Y., Hieu, L., Stefanova, L., et al.: Reverse engineering and rapid prototyping for new orthotic devices. Intell Prod Mach Syst Elsevier, pp. 567–571 (2005)

7. Todorov, G., Nikolov, N., Sofronov, Y., et al.: Creation of custom implants using 3D modeling based on CT-scan data, CAD/CAM technologies with virtual prototypes, 3D printing and high speed milling manufacturing. Comput. Methods Biomech. Biomed. Engin (2017)
8. Boboulos, M.A.: CAD-CAM & Rapid Prototyping Application Evaluation. Bookboon
9. Goh, J.C., Ho, N.C., Bose, K.: Principles and applications of computer-aided design and computer-aided manufacturing (CAD/CAM) technology in orthopaedics. Ann. Acad. Med. Singap. **19**, 706–713 (1990)
10. Hieu, L.C., Bohez, E., Vander Sloten, J. et al.: Design and manufacturing of personalized implants and standardized templates for cranioplasty applications. In: Industrial Technology, 2002. IEEE ICIT '02. 2002 IEEE International Conference on, vol.2, pp. 1025–1030 (2002)
11. Todorov, G., Savov, I., Sofronov, Y.: Possibilities for application of modern CAD/CAM/CAE to the production of orthoses. YoungFIT 79–86 (2014)
12. Todorov, G., Kamberov, K., Stoev, S., Savov, I.: Structural Analysis and Verification of Snap-Fit Joints in Plastic Parts. CONTECH 13 (2013)
13. Todorov, G., Kamberov, K., Savov, I.: Automated approach for CAD design of bone fracture orthoses. Int. Sci. Conf. 70 Years FIT, pp. 391–398 (2015)

Chapter 9
Dynamic Modeling of a Sheep Hair Shearing Device Using RecurDyn and Its Verification

Sasanka Sekhar Sinha and Subir Kumar Saha

Abstract This paper discusses a general procedure for creating a computer-aided design (CAD) model of an existing commercial sheep hair shearing device for simulation purpose. In this paper, simulation was carried out using the commercial software RecurDyn while the CAD models of the components of the device were created using SolidWorks software. Note that the mechanism in the device can be identified as an RCCR spatial four-bar linkage which was also modeled analytically from its kinematics point of view. The dynamic equations of motion were developed using the concept of the cut-joint approach and the Decoupled Natural Orthogonal Complement (DeNOC) matrices. The equations of motion were then used to simulate the mechanism in Matlab environment. The results were compared to those obtained using the CAD model in RecurDyn environment.

Keywords Sheep · Wool · Spatial · Four-bar · RCCR · RecurDyn · Lagrange multipliers

9.1 Introduction

A sheep hair shearing device is used to cut the woolen fleece of sheep. It is a power-driven machine-shear substitute for the traditional scissors to reduce the drudgery of operation. The device consists of a motor, handpiece, flexible drive, comb and cutter. The essential details of the device are shown in Fig. 9.1. From kinematic point of view, the device is a spatial four bar mechanism housed in an ergonomic chamber called the barrel shown in Fig. 9.1a. Motor provides rotary torque input to the crank by means of a flexible shaft coupling as shown in Fig. 9.1b. A fork connects the crank to a toothed blade or cutter which is driven back and forth over the surface of a comb, causing the shearing of wool. The cutter constantly presses against the

S. S. Sinha (✉) · S. K. Saha
Indian Institute of Technology Delhi, New Delhi, India
e-mail: mez168275@mech.iitd.ac.in; saha@mech.iitd.ac.in

E. Zahariev, J. Cuadrado (eds.), *IUTAM Symposium on Intelligent Multibody Systems – Dynamics, Control, Simulation*, IUTAM Bookseries 33,
https://doi.org/10.1007/978-3-030-00527-6_9

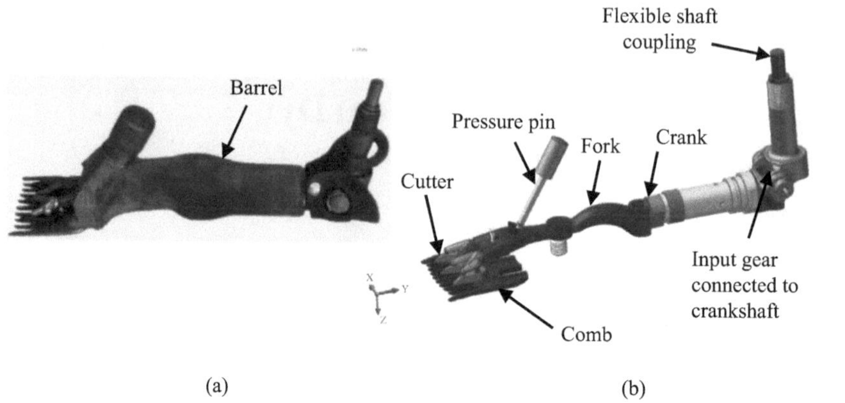

Fig. 9.1 (**a**) An actual handpiece, (**b**) CAD Model. Essential components of a sheep hair shearing device

comb and the pressure can be adjusted by means of a pressure adjustment pin. This has a direct impact on the cutting efficiency and the power consumption.

The device is manufactured by many companies in the world, e.g., Heiniger, Lister, Oster and others. Presently, it is imported at high costs in India. The Rural Technology Action Group (RuTAG) at IIT Delhi in India has been trying to develop an indigenous version since 2006. Presently, it is working on the mass scale manufacturing of the device. In order to choose appropriate materials for the components and processes to manufacture them, it was felt important that the complete dynamic analysis of an existing device be performed for deeper understanding of the interactions between the components. Therefore, a RecurDyn [1] model was attempted using the existing CAD models of the components.

In order to verify the RecurDyn results, an analytical model based on the kinematic architecture of the device which is an RCCR (revolute-cylindrical-cylindrical-revolute) spatial four-bar mechanism, was also developed. Compared to other spatial four-bar linkages like the well-known Bennett's linkage [2, 3] and the RSSR mechanisms [4] which have been studied extensively, there is not much work done on the kinematic and dynamic analysis of the RCCR mechanism to the best of the authors' knowledge. The dynamic equations of motion of the RCCR mechanism were generated using the cut-joint approach [5] and the concept of the DeNOC matrices [6]. The results were then compared to verify the outputs of RecurDyn software and the DeNOC-based dynamic model in Matlab environment.

Note that similar analyses have been reported in the literature for improved performance of a developed system. For example, Chaudhary et al. [7] developed a multiloop planar carpet scraping mechanism using a Hoeken's four-bar mechanism and a Pantograph. The cut-joint approach was used to develop the dynamic equations of motion. A comparison of the calculated driving torques was made with the values obtained from ADAMS (Automatic Dynamic Analysis of Mechanical Systems) software. Wang [8] simulated a high-speed cutting tool system in ADAMS environment and predicted any possible dynamic interference through analysis, which was later used for optimized design. Prior work on the sheep hair shearing device at IIT Delhi by Rane et al. [9] focused on the study of various kinematic and dynamic parameters, e.g., angle of fork oscillations, torque required at the crankshaft, etc. They had performed a simulation study in ADAMS but analytical expressions were derived only for the kinematics. So an analytical dynamic model was not available for comparison of the results and design optimization purpose. In this domain, one of the pioneering publications was made by Trevelyan [10] towards sensing and control of sheep-shearing robots.

The paper is organized as follows: Sect. 9.2 deals with solid modeling of various components of the sheep hair shearing device at hand. Section 9.3 discusses the model preparation and assembly details for the simulation performed in RecurDyn software. Section 9.4 delineates the simplified kinematic model chosen to replicate the behavior of the actual model. The dynamic formulation of the simplified RCCR model is also presented in this section. Lastly, Sect. 9.5 is concerned with the discussion of the results obtained from the study.

9.2 CAD Modeling of the Components

Each component of the device was scanned under a white light scanner and its shape was obtained as a point cloud of geometric samples on the surface. The measured data was devoid of any topological information. They were processed and modeled into a usable format. The level of sophistication proceeded as polygonal mesh models, NURBS surface models to editable feature-based solid models.

The first step of reconstruction to polygonal model involved finding and connecting adjacent points with straight lines in order to create a continuous surface. To do this, MeshLab was used. It is an open source, portable, and extensible system for the processing and editing of unstructured 3D triangular meshes [11]. Remeshing filters helped in surface reconstruction from points. Mesh Cleaning Filters aided removal of duplicated, unreferenced vertices, null faces, small isolated components and automatic filling of holes. MeshLab also converted .stl graphics to a normal .stl mesh file. White light scanned image and the polygonal mesh model for the comb are shown in Figs. 9.2 and 9.3, respectively.

The next step was to export into a CAD software to complete the model in CAD. Typical CAD software packages offer two basic methods for the creation of surfaces. First is the construction of curves (splines) from which the 3D surface is obtained through sweep or meshed through, and the second one is the direct creation of the surface with manipulation of the control points. Commercial CAD software SolidWorks [12] was used in the present simulation study. A combination of the geometric and freeform surfaces was used to render the mathematical identity. The end results were editable parametric CAD models for each component.

The objects and features of each created solid model were modifiable, and thereby, provided the scope for product improvement. The CAD model was then easily converted into the .step format. The G-codes can then be generated, if required, for subsequent CNC machining. The individual CAD files were assembled

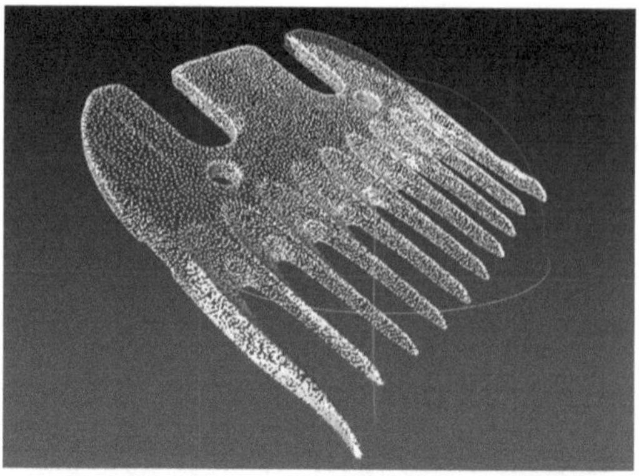

Fig. 9.2 White light scanned image of a comb

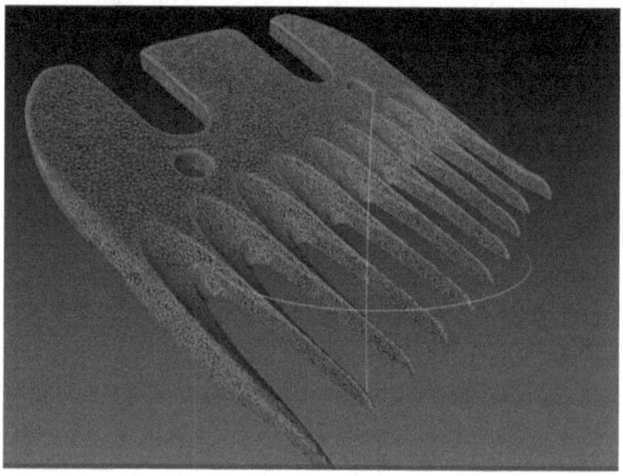

Fig. 9.3 Polygonal mesh model of the comb

together and the assembly of the handpiece is shown in Fig. 9.4. Once the CAD assembly of the handpiece was available, the sheep hair shearing device was subjected to motion analysis to check possible interference of the links during operation. Corrections were made and clearances were provided to accommodate free movement of the links by exploring the capabilities of the CAD software. It is much simpler to investigate a product in the CAD environment and incorporate changes before the real prototype is made. The model was then ready for the dynamic analysis, which can calculate the power consumption of the device for a prescribed angular velocity of the crankshaft, forces on the various components, etc.

9.3 Modeling in RecurDyn

RecurDyn has its own Modeler to create an analysis model. Alternatively, existing CAD files can also be imported in various formats. In our case, the assembly was imported in Parasolid format in RecurDyn V8R3. The joints were defined in the

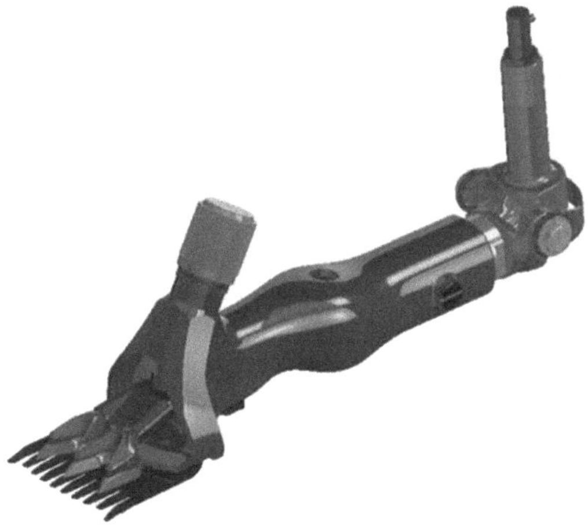

Fig. 9.4 CAD assembly of a handpiece

RecurDyn preprocessor environment. The crankshaft has one revolute joint with the barrel, which is ground for our analysis, and one cylindrical joint with the sliding ball. The fork has a cylindrical joint with the sliding ball and a revolute joint with the barrel. A spherical joint connects the pressure pin with the fork. RecurDyn is particularly efficient in handling contact problems. In the shearing device, there are quite a few places where contact modeling was required. For example, between the cutter and the comb, there is a contact which causes the rubbing action.

9.4 Inverse Dynamics

The model as shown in Fig. 9.4 is essentially a single degree-of-freedom (DoF) spatial four-bar mechanism which is shown in Figs. 9.5 and 9.6a. The joints at A and D are the revolute (R) joints with rotations θ_1 and θ_2, respectively. The joints at B and C are cylindrical (C) joints. The axes of the first two joints are parallel to each other, and so are the last two joints. The two sets of axes are perpendicular to each other. Using the cut-loop strategy for dynamic modeling as mentioned in [2], the two cylindrical joints were cut resulting in three serial open-loop subsystems. They are shown in Fig. 9.6b.

Two of the sub-systems are single-link planar serial manipulators while the third one consists of a floating body. The kinematic constraints lost due to the cut joints were recovered with the help of Lagrange multipliers. Let m_1 and m_2 be the masses, and I_{1zz} and I_{2zz} be the mass moment of inertia of the single-link subsystems. Also m_3 be the mass of the floating body of subsystem #3.The dynamic equations of motion for the first sub-system can be obtained easily as:

$$\mathbf{I}_I\ddot{\boldsymbol{\theta}}_I + \mathbf{C}_I\dot{\boldsymbol{\theta}}_I + \boldsymbol{\gamma}_I = \boldsymbol{\tau}_I + \mathbf{J}_I^T\boldsymbol{\lambda}_I \tag{9.1}$$

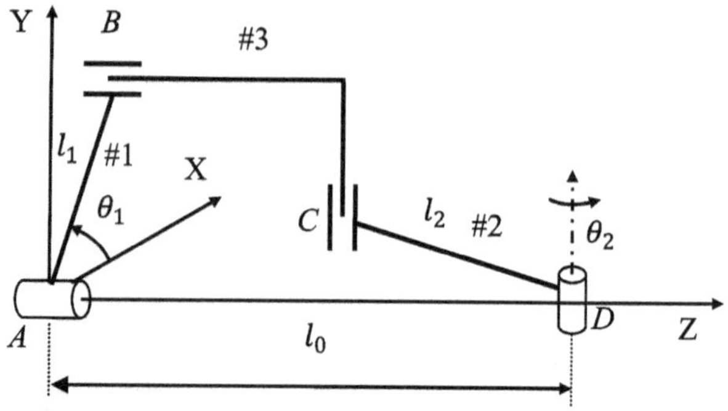

Fig. 9.5 Simplified model of sheep the hair shearing 4-bar mechanism

where the only term in the inertia matrix $\mathbf{I}_I$, i.e., scalar i_1 is $i_1 = m_1 r_1^2 + I_{1zz}$, $r_1$being the distance of the center of mass (COM) of the link from the revolute joint. The Coriolis and Centrifugal term in $\mathbf{C}_I \dot{\boldsymbol{\theta}}_I$ does not exist for the single-link manipulator. The joint torque of sub-system I is denoted as $\boldsymbol{\tau}_I = \tau_1$. The vector of Lagrange multipliers is $\lambda_I = [\lambda_{1x} \; \lambda_{1y}]^T$. It is to be noted that there is no λ_{1z} as the cylindrical joint offers no resistance in the sliding direction. The Jacobian matrix of the sub-system I is $\mathbf{J}_I = \left[-l_1 \text{sin}\theta_1 \; l_1 \text{cos}\theta_1 \right]^T$. Similarly, for sub-system II the dynamic equations of motion are written as:

$$\mathbf{I}_{II} \ddot{\boldsymbol{\theta}}_{II} + \mathbf{C}_{II} \dot{\boldsymbol{\theta}}_{II} + \boldsymbol{\gamma}_{II} = \boldsymbol{\tau}_{II} + \mathbf{J}_{II}^T \boldsymbol{\lambda}_{II} \tag{9.2}$$

where $\mathbf{I}_{\text{II}}$ also has one term, i.e., scalar $i_2 = m_2 r_2^2 + I_{2zz}$, where r_2 is the distance of the COM of the link from the revolute joint. There is no Coriolis and Centrifugal term. The joint torque of the sub-system II is 0, i.e., $\boldsymbol{\tau}_{II} = \mathbf{0}$. The vector of Lagrange multipliers is $\boldsymbol{\lambda}_{II} = \left[\lambda_{2x} \; \lambda_{2z} \right]^T$ with vanishing λ_{2y}. The coordinates of the point C can be written as $(l_2 \;\sin\theta_2, 0, l_0 + l_2 \cos\theta_2)$. Subsequently, the Jacobian matrix for the sub-system II is obtained as $\mathbf{J}_{II} = \left[-l_2 \text{sin}\theta_2 \; l_2 \text{cos}\theta_2 \right]^T$. The Newton's equations of motion for the subsystem III are then expressed as:

$$m_3 \begin{bmatrix} \ddot{x} \\ \ddot{y} \\ \ddot{z} \end{bmatrix} = - \begin{bmatrix} \lambda_{1x} + \lambda_{2x} \\ \lambda_{1y} \\ \lambda_{2z} \end{bmatrix} \tag{9.3}$$

The floating body is in rotational equilibrium. So no Euler equations of motion were considered. The coordinates of the COM of the floating body can be written as $(l_1 \cos\theta_1, l_1 \sin\theta_1 + k_1, l_2 \cos\theta_2 + k_2)$, where k_1 and k_2 are constants. The angle θ_2 can be obtained from the loop-closure constraint in the x-direction as

$$l_1 \text{cos}\theta_1 - l_2 \text{sin}\theta_2 = 0 \tag{9.4}$$

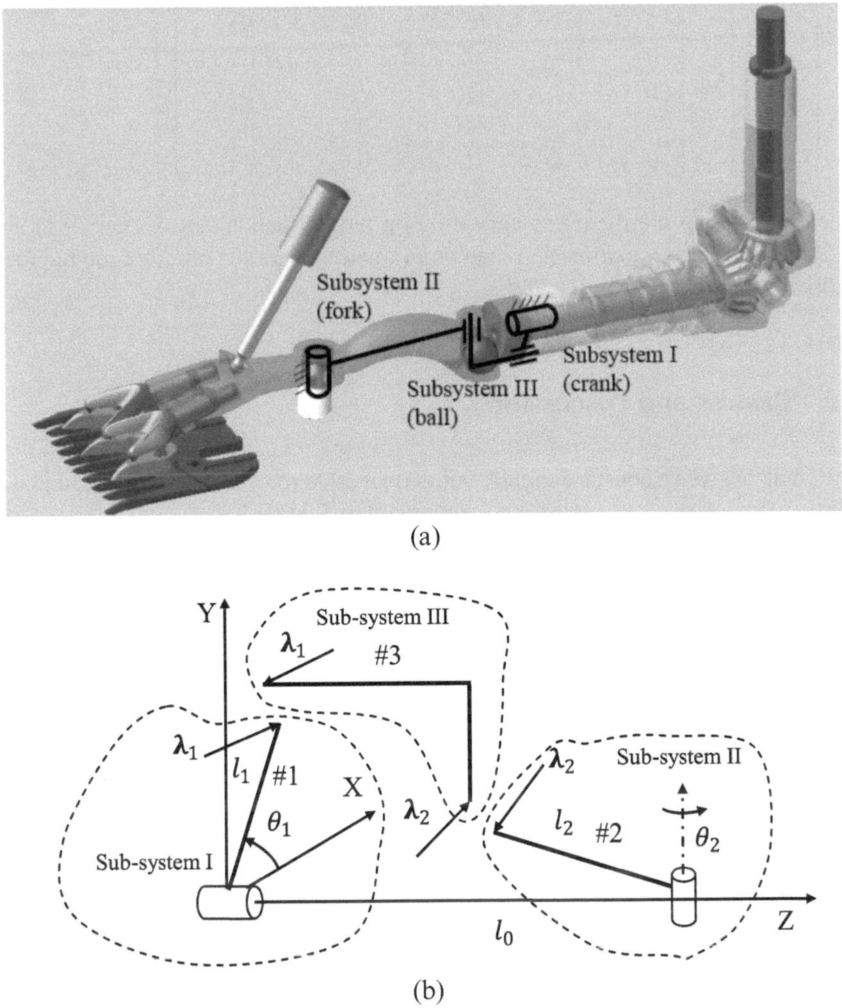

Fig. 9.6 Three cut-open subsystems

Since we have only one equation to solve for θ_2, it yields to

$$\theta_2 = \begin{cases} \pi - \sin^{-1}\left(\frac{l_1 \cos\theta_1}{l_2}\right), if \ \cos\theta_1 \geq 0 \\ \pi + \sin^{-1}\left(\frac{-l_1 \cos\theta_1}{l_2}\right), otherwise \end{cases} \tag{9.5}$$

Equations (9.1–9.3) are now rewritten in a compact form as

$$\begin{bmatrix} \varphi_I \\ \varphi_{II} \\ m_3\ddot{x} \\ m_3\ddot{y} \\ m_3\ddot{z} \end{bmatrix} = \begin{bmatrix} 1 & \mathbf{J}_I(1,1) & \mathbf{J}_I(2,1) & 0 & 0 \\ 0 & 0 & 0 & \mathbf{J}_{II}(1,1) & \mathbf{J}_{II}(2,1) \\ 0 & -1 & 0 & -1 & 0 \\ 0 & 0 & -1 & 0 & 0 \\ 0 & 0 & 0 & 0 & -1 \end{bmatrix} \begin{bmatrix} \tau_1 \\ \lambda_{1x} \\ \lambda_{1y} \\ \lambda_{2x} \\ \lambda_{2z} \end{bmatrix} \tag{9.6}$$

where φ_I and φ_{II} are the terms appearing on the left-hand sides of Eqs. (9.1) and (9.2), respectively. Equation (9.6) can be evaluated to get the input torque required to drive the crank.

9.5 Results and Discussion

The crank is given a constant angular velocity of 3000 rpm. The structural and mass-inertia parameters of the model are enumerated in Table 9.1.

Corresponding angular displacement of the fork is plotted in Fig. 9.7. The no-load torque at the crank calculated using Eq. (9.6) is plotted in Fig. 9.8. This torque is well within the specified torque limits of the motor used, i.e., 1800 Nmm. For mechanical design calculations of the components, one can account for the

Table 9.1 Structural and mass-inertia parameters for the RCCR model

Component	m_i (kg)	l_i (mm)	r_i (mm)	I_i (kgmm^2)
Crank	6.402e–2	7.794	0.5	1.639
Fork	0.115	57.8	10	280
Ball	5.06e-3	–	–	–

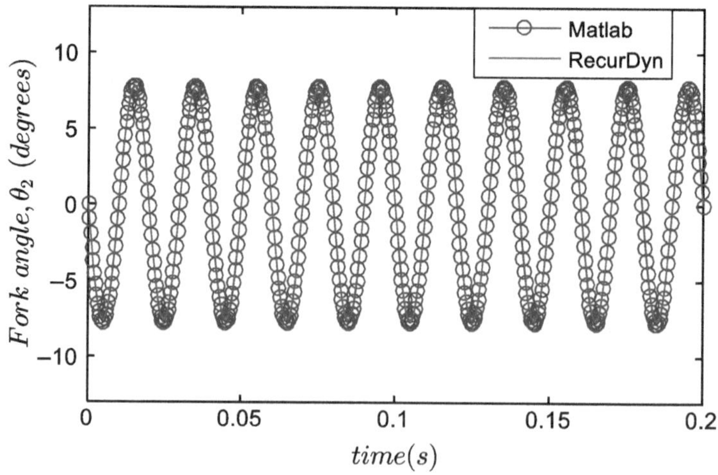

Fig. 9.7 Output motion at the fork end

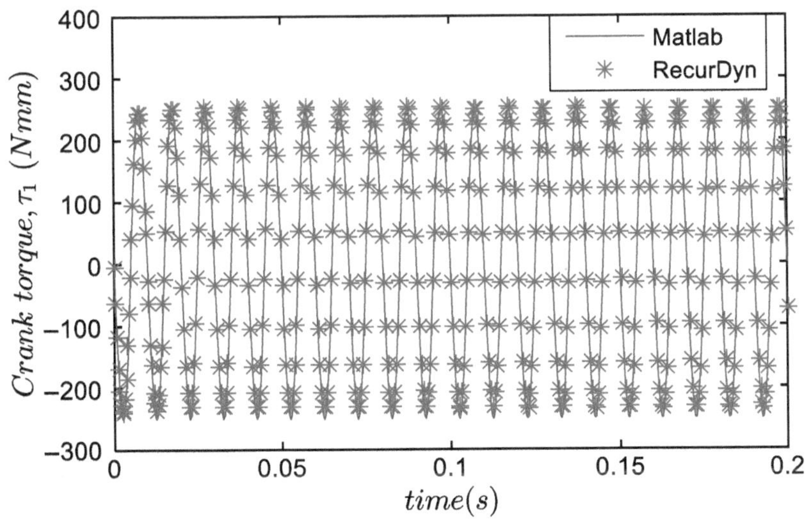

Fig. 9.8 Torque requirement at the crank

assumption of no friction considering a safety factor of 3–4 which yields the maximum required torque to be around 1000 Nmm. The fork completes one complete cycle of oscillation for every rotation of the crank leading to a gear ratio of 1.

9.6 Conclusions

The paper proposes the dynamic simulation of a sheep hair shearing device using commercial software, RecurDyn. It was also shown that the basic mechanism used for the device is a four-bar one-DoF spatial mechanism. Its dynamics modeling is also proposed using the cut-joint approach which provided the required joint-torque at the crank that is same as obtained using RecurDyn. The in-house or RecurDyn model can be easily adapted for further design improvement of the device which will be taken up in future.

Acknowledgments The work presented in this paper was carried out under Project No. RP03164 at Indian Institute of Technology (IIT) Delhi in India. It was funded by the Central Wool Development Board of India. The objective of the project was "Dissemination of low-cost sheep hair shearing device by IIT Delhi." The authors would like to thank the members of Rural Technology Action Group (RuTAG) IIT Delhi for their much-extended help and support.

References

1. RecurDyn homepage.: http://www.functionbay.org/
2. Bennett, G.T.: A new mechanism. Engineering. **76**, 777–778 (1903)
3. Hervé, J. M.; Dahan, M.: The two kinds of Bennett's mechanisms. In: Proc. Sixth IFToMM World Congress on Theory of Machines and Mechanisms, New Delhi, Vol.1, pp. 116–119 (1983)
4. Gupta, K.C., Kazerounian, S.M.: Synthesis of fully rotatable RSSR linkages. Mech. Mach. Theory. **18**(3), 199–205 (1983)
5. Chaudhary, H., Saha, S.K.: Dynamics of Closed-loop Systems in Dynamics and Balancing of Multibody Systems, SE – 3, vol. 37, pp. 45–86. Springer, Berlin/Heidelberg (2009)
6. Saha, S.K.: Dynamics of serial multibody systems using the decoupled natural orthogonal complement matrices. J. Appl. Mech. **66**(4), 986–996 (1999)
7. Chaudhary, H., Saha, S.K.: Constraint wrench formulation for closed-loop systems using two-level recursions. J. Mech. Des. **129**(12), 1234–1242 (2007)
8. Wang, S. K.: 3D motion simulation to the mechanism of 4 axis CNC machining center. Taiwan ADAMS User Seminar and Academic Thesis Delivery (2002)
9. Rane, S.; Grover, V.; Vashista, V., Jere, A., Saha, S.K.: Computer aided analysis of a sheep shearing machine. In: Proceedings of the National Conference on Emerging Trends in Mechanical Engineering, pages 1–10, SVNIT, Surat, June 4–5, (2007)
10. Trevelyan, J. P.: Sensing and control for sheep shearing robots. IEEE Transactions on Robotics and Automation 5, 6, pp. 716–727 (1989)
11. Meshlab homepage.: http://www.meshlab.net/#about
12. Solidworks homepage.: http://www.solidworks.in/

Chapter 10
Seismic Response of Soil-Structure Systems via BIEM and FEM in Absolute Coordinates

Evtim Zahariev and Petia Dineva

Abstract The motivation for this work is to model the seismic response of a structure taking into account the base three Earth components (seismic source, wave path and local soil profile) plus the engineering structure at the end of the line and modelling all of them in one system. The main aim is to develop an efficient hybrid hi-performance methodology and software that model the dynamic response of structures during earthquake accounting for the main characteristics and mechanical properties of the soil and seismic source, plus the specific structural peculiarities and mechanical behavior of the building/underground structure. The present study investigates the soil-foundation-structure interaction and the influence of the structural dynamics over the whole system's motion. The boundary integral equation method (BIEM) is applied to model the semi-infinite part of the geological domain, while the finite soil profile is described via finite element method (FEM). The structural dynamics is simulated using finite elements in absolute coordinates (FEAC), which allows the geometrical nonlinearity in dynamic behavior of the engineering structure to be taken into account. Example of forced motion of the rigid foundation as a result of wave propagation overlapped by a four stroke structural displacements illustrates the efficiency of the hybrid model.

Keywords Finite elements in absolute coordinates (FEAC) · Boundary integral equation method (BIEM) · Seismic excitations · Multibody system dynamics · Structural dynamics

E. Zahariev (✉) · P. Dineva
Institute of Mechanics, Bulgarian Academy of Sciences, Sofia, Bulgaria
e-mail: evtimvz@bas.bg; petia@imbm.bas.bg

E. Zahariev, J. Cuadrado (eds.), *IUTAM Symposium on Intelligent Multibody Systems – Dynamics, Control, Simulation*, IUTAM Bookseries 33,
https://doi.org/10.1007/978-3-030-00527-6_10

10.1 Introduction

Investigation of the dynamic interaction of engineering structures with a semi-infinite soil region due to natural (earthquakes), technical (railway traffic or pile driving) or human-made (e.g. explosions) excitations is a subject of fundamental interest in soil dynamics, civil engineering, computational mechanics, geophysics, etc. The risk analysis of structure's behavior is a typical multi-disciplinary study. An efficient way to perform this study consists in developing of high-performance models and computational tools for simulation of dynamic behavior of the soil-structure system (SSS). Earthquake engineering as a scientific discipline is therefore a key component underpinning the activities of the construction industry operating in earthquake-prone regions and of the various agencies responsible with regulatory issues and with post-earthquake relief and rehabilitation efforts. A brief state-of the-art will be presented, focusing on seismic wave propagation in complex geological media and also on seismic response of the whole soil-structure system. Essentially, three main groups of approaches treat the problem and they are analytical, numerical and hybrid.

The known analytical methods are ray theory [1], generalized ray technique, mode matching methods [2], reflectivity method [3, 4], wave number integration method [5] and others, all restricted to media with simple geometry, with inhomogeneity dimensions that are considerably larger than the prevailing wavelengths, and therefore are only of limited use for seismic zoning and micro-zoning studies.

On the opposite side, the numerical techniques as finite difference method (FDM) [6], finite element method (FEM) [7, 8], boundary integral equation method (BIEM) [9–11] are suitable for studying structures with complex geometry and complex mechanical properties.

The BIEM has proven to be an accurate numerical method for problems in elastodynamics, see [11–13]. This technique is based on a transformation of the governing partial differential equations, which involve unknown fields inside and on the boundary of a domain, into a system of integral equations along existing boundaries. The BIEM requires two basic ingredients: a reciprocal relation and a fundamental solution of the governing equation. Combined with the boundary element method (BEM) the fundamental solution (or Green's function) provides an elegant and yet powerful tool in investigating various complicated responses of geomaterials/civil structures.

Some of the main advantages of the BIEM in treating seismic wave propagation are: (a) reduction in the dimension of the problem; (b) semi-analytical character of the method; (c) high level of accuracy since numerical quadrature techniques are directly applied to the boundary integral equations; (d) BEM has proven to be powerful tool for solving exterior boundary-value problems and wave propagation problems in multilayered geological regions, because only the boundaries between layers are discretized; (e) the fundamental solution used in the construction of the boundary integral equations obeys the radiation condition, and thus, infinitely extended boundaries are automatically incorporated, in contrast to other numerical methods where special transmitting and/or non-reflecting viscous boundaries have to be used.

When modelling seismic wave propagation, the three basic components of the problem are the source, the travel path, and the receiving site. Soil-structure interaction (SSI) problem is one of the most classical problems connecting two different research fields as wave propagation in continua and dynamic response of structures. Once the seismic waves hit the structure, the input motion excites the structure and simultaneously it gets modified by the movement of the latter relative to the ground. The behavior of the structure and the surrounding soil are interconnected and the analysis of both in a compatible manner is necessary.

The computational technique proposed here aims to simulate the dynamic response of structures embedded in an elastic half-space basing on the hybrid approach combining different type of computational tools [14, 15]. The study presents an efficient hybrid hi-performance methodology and software that model the dynamic response of structures during earthquake accounting for the main characteristics and mechanical properties of the soil and of the seismic source, plus the specific structural peculiarities and mechanical behavior of the building/underground structure construction.

The inventive methods for earthquake structure response analysis and design are based on precise dynamic modeling and simulation of multibody systems of rigid and flexible bodies. The initial conditions and the external disturbances are the initial data which estimation is of major importance for the simulation process. The earthquake wave propagation is well studied and sufficient data collection is at the disposal of the scientists and engineers. However, information about the structure foundation response because of ground motion and wave propagation is still challenging subject for the investigations.

In this work the ground is considered as environment with specific elastic and damping properties, which in certain conditions could experience plastic displacements. In Fig. 10.1 the two-dimensional kinematic model of ground – foundation interaction is shown. The ground is presented by finite element (FE) discretization (Fig. 10.1a. The structure is compiled of rigid body foundation and a flexible tower of elastic finite elements. The precise dynamic model should consider the wave propagation of the ground and the interaction with the foundation. Using the methodology of the soil-structure interaction the reduced forces (F_1 and F_2), Fig. 10.1b loading the foundation could be calculated and included in the dynamic equations of the structure. The major difficulty here is the fact that the foundation is moving (coordinates q_1, q_2) and its dynamic characteristics should be taken into account. In this case the dynamics of the entire system soil – foundation – structure is to be simulated. Implementation of this method requires enormous statistical data for the ground properties and local earthquake wave propagation.

In structural engineering passive, active and semi-active methods for structural displacement suppression are applied. In this work passive methods are considered and simulated for estimation of the proper damper characteristics – damping and spring coefficients. A simplified model for ground – foundation interaction

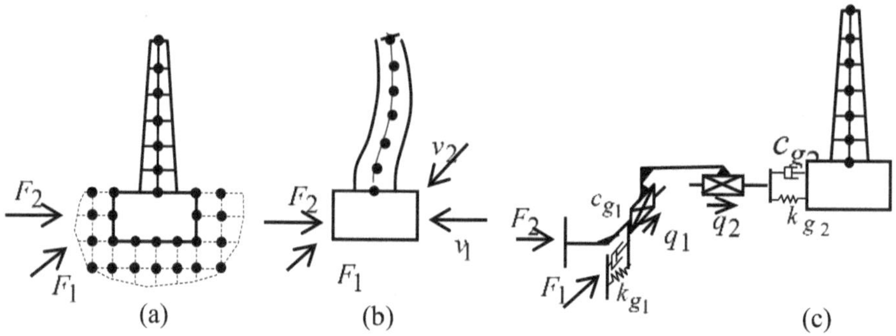

Fig. 10.1 Two – dimensional kinematic model of ground – foundation interaction

including damper properties is presented in Fig. 10.1c. Virtual springs and dampers represent the elasticity and the damping properties of the ground. The spring properties are presented as a resistant force that depends on the displacements. Since the ground could exhibit residual displacements (virtual plastic displacements) the structure could change its location.

The simulation of the foundation motion is much simpler when it is built on a rigid ground, for example rock. Then the foundation motion could be considered coincident with the ground motion read from the statistical data. This case was investigated in [16] using differential algebraic equations with reonomic constraints.

Recently a novel method of absolute nodal coordinate formulation has been developed [17, 18]. It is based on the theory of the curvilinear coordinates. The slopes of the nodes in the tangential line or plane of the space curve or shell are applied as coordinates. The method proposes important advantages mainly in large rotation simulation and mass matrices formulation. Review of the scientific literature regarding this method for dynamics simulation of large deformation of flexible multibody systems was presented by Gerstmayr et al. [19]. The paper provided a comprehensive review of the advantages and further developments.

A method of finite elements in absolute coordinates (FEAC) is discussed here for modelling of the structural dynamics. It was recently presented in [20] and further developed for different applications. The method is proposed here for simulation of structures subjected to seismic excitations. Acceleration and force methods are used in practice for dynamics simulation of the earthquake structure response. Acceleration approach consists in registration of the motion of a structure foundation by accelerographs and normally consists in three orthogonal components of the ground acceleration. The velocity and displacement of the ground are obtained integrating the data of the accelerograms. The statistical data for the specific regions are than applied as input data for the dynamic simulation. In [21] the accelerations of the foundation are reonomic constraints for the dynamic equations which are transformed with respect to the parameters that consist of structure displacement and the forces, applied to the foundation.

The present study extends the investigations on the more complicated case including the dynamics of the soil-foundation-structure interaction and the influence of the structural dynamics over the whole global system's dynamics. An example of the influence of the forced motion of the foundation as a result of soil wave propagation overlapped by a four stroke structural displacements is presented.

The main advantage of the proposed hybrid methodology is the capability to describe the influence of all components of the system: seismic source-wave path-foundation-engineering structure over the seismic response of the object under study. In the whole model is taken into account the specific geophysical characteristics and location of the seismic source, the inhomogeneity and heterogeneity of the wave path from the source till the local region containing a structure, geotechnical properties of the local finite region and finally the engineering structure itself. The BIEM model of the semi-infinite part of the geological domain is developed, while the finite geological profile is described via FEM. The structural dynamics is simulated using the computational technique FEAC [16], which allows the geometrical nonlinearity in seismic behavior of engineering structure to be taken into account..

10.1.1 Mechanical Problem Formulation

The system under consideration consists of the following subdomains, (Fig. 10.2):

- Semi-infinite soil domain Ω_0 **with** boundary $\Gamma_0 = \infty A \bigcup AB \bigcup BC \bigcup CD \subset D\infty$ and free surface $\Gamma_{00} = \infty A \bigcup D\infty$.
- Finite domain of sand layer Ω_1 with boundary $\Gamma_1 = AB \bigcup BC \bigcup CD \bigcup DN \bigcup NQ \bigcup QP \bigcup PM \bigcup MA$ and free surface $\Gamma_{10} = MA \bigcup DN$. This layer acts as a damper of the seismic energy before its entering in the foundation. The interface boundary between passive isolation layer Ω_1 and geological soil half-plane Ω_0 is $AB \bigcup BC \bigcup CD$, while between damped layer Ω_1 and the foundation Ω_f is $NQ \bigcup QP \bigcup PM$.
- Rigid body foundation Ω_f with boundary $\Gamma_f = MP \bigcup PQ \bigcup QN \bigcup NM$ and common boundary with the building structure Ω_s denoted by NM.
- Elastic building structure Ω_s with boundary $\Gamma_s = MN \bigcup NF \bigcup FT \bigcup TM$.

The whole system is subjected to an incident transient wave radiated from a seismic source located at point $X_0(x_{01}, x_{02})$ in the geological region Ω_0 assumed to be unaffected by the presence of the soil layer Ω_1, the foundation Ω_f and the structure Ω_s.

Plane strain state and respectively in-plane wave motion in the plane $x_3 = 0$ of a Cartesian coordinate system $Ox_1x_2x_3$ is assumed. The only non-zero field quantities are displacement components u_1, u_2, stresses $\sigma_{11}, \sigma_{22}, \sigma_{12}$ and tractions t_1, t_2 dependent on the coordinates $x(x_1, x_2)$ and on the time t.

The aim is to describe the seismic behaviour of the whole system: soil region-rigid foundation-building structure via efficient hybrid methodology. The approach

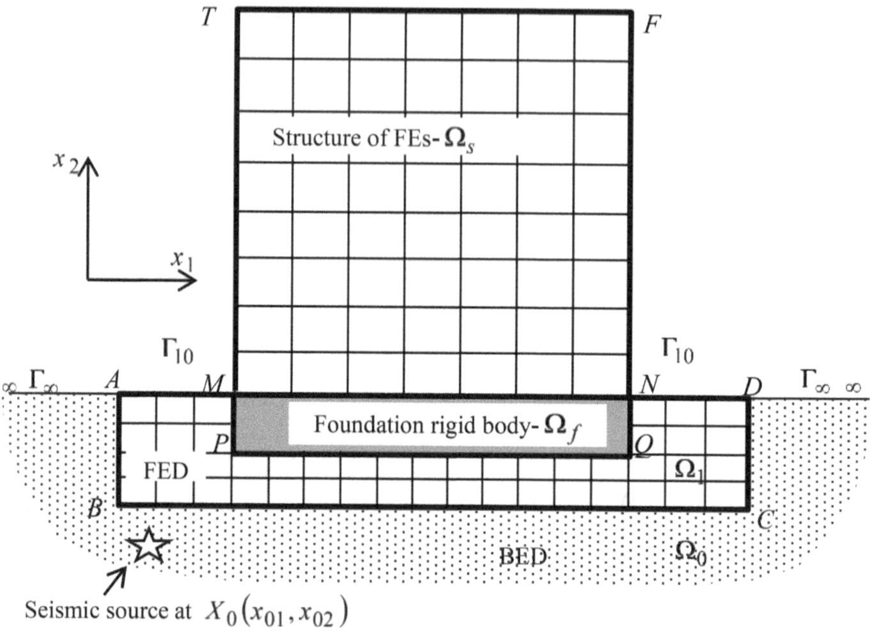

Fig. 10.2 The geometry of the soil-foundation-building structure

is based on the boundary integral equation method for describing the wave field in the semi-infinite soil region Ω_0 and on the finite element method for evaluating the seismic response of the finite bounded sand layer Ω_1 which contains a building structure $\boldsymbol{\Omega}_s$ resting on the rigid foundation $\boldsymbol{\Omega}_f$. The FEAC allowing to model the geometrical nonlinearity in dynamic behavior of the engineering structure is applied.

10.2 Boundary Value Problem Formulation

The dynamic behaviour of the soil-foundation-structure system in Fig. 10.2 is described by the equation of motion in transient elastodynamics:

$$\sigma_{ij,j}^{k}(x,t) = \rho^{k}\frac{\partial^2 u_i^k(x,t)}{\partial t^2}, \quad x(x_1,x_2) \in \Omega_k, \; k = 1, s, \tag{10.1}$$

$$\sigma_{ij,j}^{0}(x,t) + f_i(X_0,t)\,\delta(x - X_0) = \rho^{0}\frac{\partial^2 u_i^0(x,t)}{\partial t^2}, \quad x(x_1,x_2) \in \Omega_0, \tag{10.2}$$

where: ρ^0, ρ^1 and ρ^s are the densities of the elastic domains Ω_0, Ω_1 and Ωs; $x(x_1,x_2)$ are the coordinates of the observer point; $X_0(x_{01},x_{02})\in\Omega_0$ are the co-ordinates of the point where the seismic source is located; $u_i^k\,(x,t)$ and $\partial^2 u_i^k/\partial t^2$ are the components of displacement and acceleration vectors in the k-th domain; $\sigma_{ij}^k\,(x,t)$ is the elastic stress tensor in the k-th domain; δ is Dirac's delta function. The seismic concentrated force is $f_i(x_0,t) = f_{0i}f(t)$ with magnitude $f_{0i}(f_{01},f_{02})$ and time function $f(t)$. In Eqs. (10.1 and 10.2) summation convention over repeated indices is applied and a subscript comma denotes the partial differentiation.

In order to formulate the initial boundary value problem (BVP) we should add the initial and boundary conditions. Note that the continuum is assumed to be at rest before time $t = 0$, so zero initial conditions $u_i = 0$; $\partial u_i/\partial t = 0$, $t < 0$ hold.

The solutions of the problem for transient waves is solved by the use of the following well known numerical procedures:

(a) The direct Fourier transform is applied to the governing equations, Eqs. 10.1 and 10.2, and the corresponding BVP in frequency domain with governing equations, Eqs. 10.3, 10.4, is solved;
(b) The inverse FFT (fast Fourier transform) is applied to the solutions in frequency domain and finally solutions in time domain are determined.

After application of the direct Fourier transform in respect to the time variable, the hyperbolic partial differential equations, Eqs. 10.1 and 10.2, transform to the elliptic ones in the frequency range:

$$\sigma_{ij,j}^k\,(x,\omega) + \rho^k\omega^2 u_i^k\,(x,\omega) = 0,\ \ x\,(x_1,x_2) \in \Omega_k,\ \ k = 1,s, \tag{10.3}$$

$$\sigma_{ij,j}^0\,(x,\omega) + \rho^0\omega^2 u_i^0\,(x,\omega) = -f_{0i}\,\widehat{f}\,(\omega)\ \ \delta\,(x - X_0)\,,\ \ x\,(x_1,x_2) \in \Omega_0, \tag{10.4}$$

In Eq. 10.4 $\widehat{f}\,(\omega)$ is the Fourier transform of the time function of the seismic signal. The field variables in all domains under consideration should satisfy the boundary conditions in the frequency domain.

10.2.1 Boundary Conditions Along the Boundaries of the Semi–infinite Domain Ω_0

$$t_i^0\,(x,\omega) = 0\ \ for\ x \in \infty A \cup \infty D, \tag{10.5}$$

$$t_i^0\,(x,\omega) = -t_i^1\,(x,\omega)\ \ for\ x \in AB \cup BC \cup CD \tag{10.6}$$

$$u_i^0(x,\omega) = u_i^1(x,\omega)\ \ for\ x \in AB \cup BC \cup CD \tag{10.7}$$

At infinite $x \to \infty$Sommerfeld's radiation condition is satisfied.

10.2.2 Boundary Conditions Along the Boundaries of the Finite domain Ω_1 Consisting the Rigid Foundation Ω_f

$$t_i^0(x,\omega) = 0\ for\ x \in MA \cup DN. \tag{10.8}$$

Conditions, Eqs. 10.6 and 10.7, are satisfied for $x \in AB \bigcup BC \bigcup CD$.

For boundary $MP \bigcup PQ \bigcup QN$ between the rigid foundation Ω_f and the soil layer Ω_1 the following displacement compatibility conditions [22] hold:

$$\mathbf{u}^{k-th\ node(1)}\left(\mathbf{x}^{k-th\ node}, \omega\right) = \mathbf{C}^i \mathbf{u}^{ref}\left(\mathrm{x}^{k-th\ node}, \omega\right); \mathbf{x} \in MP \cup PQ \cup QN. \tag{10.9}$$

Here: $\mathbf{u}^{k-th\ node(1)}$ is 3D displacement vector at the k-th node $\left[x_1^{k-th\ node}\ x_2^{k-th\ node}\ x_3^{k-th\ node}\right]^{\mathrm{T}}$ of the discretization mesh along the interface boundary $MP \bigcup PQ \bigcup QN$ from the side of the elastic soil layer Ω_1; $\mathbf{u}^{ref} = \left[u_1^{ref}\ u_2^{ref}\ u_3^{ref}\ \omega_1^{ref}\ \omega_2^{ref}\ \omega_3^{ref}\right]^{\mathrm{T}}$ is 3D generalized displacement vector of the rigid body measured at an arbitrary point of reference with coordinates $\left(x_1^{ref}, x_2^{ref}, x_3^{ref}\right) \in \Omega_f$, where u_i^{ref} and ω_i^{ref} are the components of the displacement and rotation; the superscript "T" denotes matrix transpose; the matrix $\mathbf{C}^{k-th\ node}$ is as follows:

$$\mathbf{C}^{k-th\ node} = \left[\begin{array}{ccc|c|c|c} 1 & 0 & 0 & 0 & \left(x_3^{ref} - x_3^{k-th\ node}\right) & \left(x_2^{ref} - x_2^{k-th\ node}\right) \\ \hline 0 & 1 & 0 & \left(x_3^{ref} - x_3^{k-th\ node}\right) & 0 & \left(x_1^{ref} - x_1^{k-th\ node}\right) \\ \hline 0 & 0 & 1 & \left(x_2^{ref} - x_2^{k-th\ node}\right) & \left(x_1^{ref} - x_1^{k-th\ node}\right) & 0 \end{array}\right] \tag{10.10}$$

The kinematic compatibility relationship for all n_r nodes along the rigid boundary can be written as:

$$\mathbf{u}_r = \mathbf{C}\mathbf{u}^{ref}, \tag{10.11}$$

where

$$\mathbf{u}_r = \left[\mathbf{u}^{1-st\ node}\ \mathbf{u}^{2-st\ node}\ \ldots\ \mathbf{u}^{n_r-st\ node}\right]^{\mathrm{T}}, \tag{10.12}$$

$$\mathbf{C} = \left[\mathbf{C}^{1-st\ node}\ \mathbf{C}^{2-st\ node}\ \ldots\ \mathbf{C}^{n_r-st\ node}\right]^{\mathrm{T}}. \tag{10.13}$$

Equilibrium between the forces acting on the rigid foundation, and the tractions acting along the interface boundary $MP\bigcup PQ\bigcup QN$ from the elastic domain Ω_1 is required. Note that vectors and matrices are denoted by bold style or are presented by their components.

Let $\mathbf{t}^{j-th\ element} = \left[t_1^{j-th\ element}\ t_2^{j-th\ element}\ t_3^{j-th\ element}\right]^{\mathrm{T}}$ be the vector of tractions developed in the elastic layer Ω_1 along $MP\bigcup PQ\bigcup QN$ and $\mathbf{n}^{j-th\ element} = \left[n_1^{j-th\ element}\ n_2^{j-th\ element}\ n_3^{j-th\ element}\right]^{\mathrm{T}}$ be the vector of normal of the j-th finite element over the rigid boundary $MP\bigcup PQ\bigcup QN$.

Denote by $\mathbf{F}^{cq} = \left\lfloor F_1^{cg}\ F_2^{cg}\ F_3^{cg}\ M_1^{cg}\ M_2^{cg}\ M_3^{cg}\right\rfloor^{\mathrm{T}}$ the external force acting at the centre of mass of the rigid foundation Ω_f with coordinates $\left(x_1^{cg}, x_2^{cg}, x_3^{cg}\right)$ where F_i^{cg} $(i = 1, 2, 3)$ are the components of the forces and M_i^{cg} their moments.

The equilibrium relations at the centre of gravity of the rigid body foundation Ω_f can be expressed as:

$$F_k^{cg} = \sum\nolimits_{j=1}^{n_e} \int_{\Gamma_j} t_k^j \left(x_1^j, x_2^j, x_3^j, \omega\right) d\Gamma_j - \omega^2 M u_k^{cg},\ k = 1, 2, 3,\ j = 1, 2, \ldots, n_e, \tag{10.14}$$

$$M_1^{cg} = \sum\nolimits_{j=1}^{n_e} \int_{\Gamma_j} t_2^j \left(x_3^{cg} - x_3^j\right) d\Gamma_j + \sum\nolimits_{j=1}^{n_e} \int_{\Gamma_j} t_3^j \left(x_2^{cg} - x_2^j\right) d\Gamma_j - \omega^2 I_1^{cg} \phi_1^{cg}, \tag{10.15}$$

$$M_2^{cg} = \sum\nolimits_{j=1}^{n_e} \int_{\Gamma_j} t_1^j \left(x_3^{cg} - x_3^j\right) d\Gamma_j + \sum\nolimits_{j=1}^{n_e} \int_{\Gamma_j} t_3^j \left(x_1^{cg} - x_1^j\right) d\Gamma_j - \omega^2 I_2^{cg} \phi_2^{cg}, \tag{10.16}$$

$$M_3^{cg} = \sum\nolimits_{j=1}^{n_e} \int_{\Gamma_j} t_1^j \left(x_2^{cg} - x_2^j\right) d\Gamma_j + \sum\nolimits_{j=1}^{n_e} \int_{\Gamma_j} t_2^j \left(x_1^{cg} - x_1^j\right) d\Gamma_j - \omega^2 I_3^{cg} \phi_3^{cg}, \tag{10.17}$$

Here: M is the total foundation's mass, I_k^{cg} is the k-th components of the inertia moments at the centre of mass of the foundation Ω_f, n_e is the number of elements along the rigid boundary between the foundation and soil layer Ω_1, is x_k^j is the k-th coordinate of the point over the j-th finite element.

In terms of the all n_r nodal values of the field variables, the set of equations, Eqs. 10.14, 10.15, 10.16, and 10.17 may be expressed in a matrix form as:

$$\mathbf{F}^{cg} = \mathbf{E}\mathbf{t}_r - \omega^2 \mathbf{M}\mathbf{u}^{cg}, \tag{10.18}$$

Where $\mathbf{E}$ is the matrix formed by integrals involving the corresponding elemental shape functions, $\mathbf{M}$ is a diagonal matrix containing the total mass and the inertia moments of the foundation, $\mathbf{t}_r = \left[t^1\ t^2\ t^3\ \ldots\ t^{n_r}\right]^{\mathrm{T}}$ is the vector with tractions in all n_r nodes, $\mathbf{u}^{cg} = \left[u_1^{cg}\ u_2^{cg}\ u_3^{cg}\ \phi_1^{cg}\ \phi_2^{cg}\ \phi_3^{cg}\right]^{\mathrm{T}}$ is the vector of displacement of the center of gravity of the foundation.

Equation 10.18 defined at the center of gravity of the rigid body, can be generalized for any arbitrary point of reference considering equilibrium and kinematic relations between this point and the center of mass. These relations may be expressed in matrix form as follows:

$$\mathbf{F}^{cg} = \mathbf{T}\mathbf{F}^{ref};\ \mathbf{u}^{cg} = \mathbf{L}\mathbf{u}^{ref}, \tag{10.19}$$

where

$$\mathbf{T} = \left(\mathbf{L}^{\mathrm{T}}\right)^{-1}, \mathbf{L} = \begin{bmatrix} 1\,0\,0 & 0 & \left(x_3^{cg} - x_3^{ref}\right) & \left(x_2^{ref} - x_2^{cg}\right) \\ 0\,1\,0 & \left(x_3^{ref} - x_3^{cg}\right) & 0 & \left(x_1^{cg} - x_1^{ref}\right) \\ 0\,0\,1 & \left(x_2^{cg} - x_2^{ref}\right) & \left(x_1^{ref} - x_1^{cg}\right) & 0 \\ 0\,0\,0 & 1 & 0 & 0 \\ 0\,0\,0 & 0 & 1 & 0 \\ 0\,0\,0 & 0 & 0 & 1 \end{bmatrix} \tag{10.20}$$

and the superscript "-1" denotes inverse matrix. Although the boundary conditions along the boundary of the rigid foundation are presented for 3D case, their 2D version is applied in the 2D hybrid model proposed here.

10.2.3 Boundary Conditions Along the Boundaries of the Elastic Structure Ωs

The following conditions are satisfied:

(a) Conditions for zero tractions along the free surfaces $MT \bigcup TF \bigcup FN$;
(b) Continuity conditions for displacements and equilibrium conditions for forces along the common boundary MN with the rigid foundation Ω_f.

Finally, the BVP is defined in frequency domain by the governing equations, Eqs. 10.3 and 10.4, and by the discussed above boundary conditions along the existing boundaries. The solution in the time domain will be solved via application of the inverse FFT of the obtained solutions in frequency domain.

The solution of the formulated BVP will give displacements and traction at any observer point in the system under consideration.

The synthetic seismic signals depend on the type and properties of the seismic load, on the material and geometric characteristics of the whole wave path starting from the seismic source till the observer point.

10.3 Hybrid Computational Model Based on the BIEM and FEM

An efficient 2D hybrid computational tool based on the BIEM and FEM is developed. The half-plane is divided into two regions: (a) semi-infinite homogeneous far-field geological region Ω_0 containing an embedded line seismic source at a point $X_0(x_{01}, x_{02})$. This region is described by the BIEM and is named boundary element domain (BED) zone; (b) finite bounded region Ω_1 comprising a building structure Ω_s rested on rigid foundation Ω_f. This region is described by the FEM and is named finite element domain (FED) zone.

The essence of the coupling numerical methods is to best exploit the features of the constituent parts. Each of the two techniques here is applied in that part of the whole model where it works more efficiently:

(a) BIEM formulation of the elastodynamic problem is mathematically equivalent to the original formulation via partial differential equations and the fact that it is based on the fundamental solution of the governing equation, guarantees a high level of solution accuracy. Additionally the BIEM numerical scheme fits well to external problems, because the fundamental solution obeys the Sommerfeld's radiation condition and thus infinitely extended boundaries are automatically accounted for without resorting to special types of viscous boundaries as is in FEM;
(b) It is well known in structural analysis that the major strength of the FEM is its versatility in handling various classes of problems, including nonlinearity, anisotropy and nonhomogeneity and an important fact also is that there exist commercial FEM codes that can be used successfully in mechanical modelling of complex engineering structures.

The hybrid numerical scheme consists of the following steps:

(1) Direct BIEM is applied for the BED zone based on the full space elastodynamic fundamental solution in Fourier domain for unbounded geological region;
(2) Conventional FEM is used for the near soil region Ω_1 with rigid foundation;
(3) FEAC is applied in order to describe the dynamics of the building's structure;
(4) Insertion of the BIEM model of the seismically active far-field soil region as a macro finite element in the FEM model;

(5) Solution of the FEM model in frequency domain.
(6) Inverse FFT is applied to the solutions in frequency domain and finally solution in time–domain is obtained.

10.3.1 BIEM Model of the Seismic Wave Field in the Semi–infinite Domain Ω_0

The seismic wave field in the semi-infinite region Ω_0 is described via boundary integral equation (BIE) along the boundary $\Gamma_0 = \infty A \bigcup AB \subset BC \bigcup CD \bigcup D\infty$ [11], i.e.:

$$
\begin{aligned}
c_{lk} u_k^{(\Omega_0)}(x,\omega) &= \int\limits_{\Gamma_{\Omega_0}} U_{lk}^{(\Omega_0)}(x,\xi,\omega)\ t_k^{(\Omega_0)}(\xi,\omega)\ d\Gamma \\
&- - \int\limits_{\Gamma_{\Omega_0}} P_{lk}^{(\Omega_0)}(x,\xi,\omega)\ u_k^{(\Omega_0)}(\xi,\omega)\ d\Gamma \\
&\qquad + f_{0k}\hat{f}(\omega)\,U_{lk}^{(\Omega_0)}(x,X_0,\omega)
\end{aligned}
\tag{10.21}
$$

Here c_{lk} is the jump term depending on the local geometry at the collocation point x, (x,ξ) are the position vectors of the source and field points, U_{lk} is the displacement fundamental solution of (Eq. 10.4) given in [23], $P_{ij} = C_{iqls} U_{sj},{}_l n_q$ is its corresponding traction, $C_{ijkl} = \lambda\delta_{ij}\delta_{kl} + \mu(\delta_{ik}\delta_{jl} + \delta_{il}\delta_{jk})$ is the tensor of elastic constants, reducing to two Lame constants μ and λ for elastic isotropic material, δ_{ij} is Kronecker symbol.

Equation 10.21, traction free boundary condition, Eq. 10.5, along the boundary $\infty A \bigcup \infty D$, compatibility and dynamic equilibrium boundary conditions for the total displacements, Eq. 10.6, and tractions, Eq. 10.7, along the boundary $AB \bigcup BC \bigcup CD$ and Sommerfeld's radiation condition at infinite describes the BIEM model of the seismic field in the semi-infinite domain Ω_0.

A standard numerical technique based on the discretization of the BIEs by isoparametric quadratic boundary elements and the collocation procedure [11] is applied. By keeping in mind the asymptotic behaviour of the frequency-dependent fundamental solution for small arguments, two types of singular integrals appear:

(a) $\int\limits_{\Gamma_{\Omega_0}} U_{lk}(x,\xi,\omega)\ t(\xi,\omega)\,d\Gamma$ with weak singularity of order O(ln$r$) and
(b) $\int\limits_{\Gamma_{\Omega_0}} P_{lk}(x,\xi,\omega)\ u_k(\xi,\omega)\ d\Gamma$ with a strong singularity of order O(1/r), where r is the distance between the source and field points.

The weakly singular displacement kernel can be numerically integrated by special-purpose Gaussian quadrature, supplemented by a breakdown of the singular boundary element (BE) into sub-elements for better accuracy. The strongly singular traction kernel is integrated indirectly by applying the rigid-body motion concept.

The BIE, Eq. 10.21, transforms into the following matrix equation after discretization and collocation numerical procedure:

$$\mathbf{Gt} - \mathbf{Hu} = \mathbf{\Phi} \tag{10.22}$$

Here $\mathbf{H}$ and $\mathbf{G}$ are matrices of size $2N \times 2N$, N is the total number of nodes along $\Gamma_0 = \infty A \bigcup AB \bigcup BC \bigcup CD \bigcup D\infty$, $\mathbf{u}$ and $\mathbf{t}$ are displacement and traction vectors at the nodes, the vector Φ is derived through the contribution of the seismic load presented by the term $f_{0k}\widehat{f}(\omega) \cdot U_{lk}(x, X_0, \omega)$.

10.3.2 Presentation of the BIEM Model of the Semi–infinite Domain Ω_0 as a Macro–finite Element and Its Import in the FEM Model

Here DOFs condensation at the nodes along the interface boundary $AB \bigcup BC \bigcup CD$ between the semi-infinite BEM zone and finite FEM zone is discussed. The condensation is performed similar to those proposed in [14]. Let us refer the interface boundary $AB \bigcup BC \bigcup CD$ to index 1 while the free surface boundary $\Gamma_{00} = \infty A \bigcup D\infty$ is referred to index 2. Then Eq. 10.22 can be rewritten as follows:

$$\begin{bmatrix} \mathbf{G}_{11} & \mathbf{G}_{12} \\ \mathbf{G}_{21} & \mathbf{G}_{22} \end{bmatrix} \begin{bmatrix} \mathbf{t}_1 \\ \mathbf{t}_2 \end{bmatrix} - \begin{bmatrix} \mathbf{H}_{11} & \mathbf{H}_{12} \\ \mathbf{H}_{21} & \mathbf{H}_{22} \end{bmatrix} \begin{bmatrix} \mathbf{u}_1 \\ \mathbf{u}_2 \end{bmatrix} = \begin{bmatrix} \mathbf{\Phi}_1 \\ \mathbf{\Phi}_2 \end{bmatrix}. \tag{10.23}$$

From the second row we can write for $\mathbf{u}_2$:

$$\mathbf{u}_2 = \mathbf{A}_t \mathbf{t}_1 + \mathbf{A}_u \mathbf{u}_1 + \mathbf{\Theta}, \tag{10.24}$$

where: $\mathbf{\Theta} = \mathrm{H}_{22}^{-1}(\mathbf{G}_{22}\mathbf{t}_2 - \mathbf{\Phi}_2)$, $\mathbf{A}_t = \mathrm{H}_{22}^{-1} \cdot \mathbf{G}_{21}$ and $\mathbf{A}_u = \mathrm{H}_{22}^{-1} \cdot \mathbf{H}_{21}$. After inserting Eq. 10.24 in the first row of Eq. 10.23 we obtain:

$$(\mathbf{G}_{11} - \mathbf{H}_{12}\mathbf{A}_t)\ \mathbf{t}_1 = (\mathbf{H}_{12}\mathbf{A}_u + \mathbf{H}_{11})\ \mathbf{u}_1 + \mathbf{H}_{12}\mathbf{\Theta} - \mathbf{G}_{12}\mathbf{t}_2 + \mathbf{\Phi}_1 \tag{10.25}$$

Having in mind that $\mathbf{t}_2(x \in \Gamma_{00}, \omega) = 0$, the following compact relation between tractions t_1 and displacements u_1 along the contact interface zone $AB \bigcup BC \bigcup CD$ is obtained:

$$\mathbf{t}_1 = \mathbf{B}\mathbf{u}_1 + \mathbf{P}, \tag{10.26}$$

where $\mathbf{B} = [\mathbf{G}_{11} - \mathbf{H}_{12}\mathbf{A}_t]^{-1}(\mathbf{H}_{12}\mathbf{A}_u + \mathbf{H}_{11})$; $\mathbf{P} = [\mathbf{G}_{11} - \mathbf{H}_{12}\mathbf{A}_t]^{-1}$ $(\mathbf{H}_{12}\mathbf{\Theta} - \mathbf{G}_{12}\mathbf{t}_2 + \mathbf{\Phi}_1)$.

As can be seen from Eq. 10.26 the relation between traction and displacement along the contact zone $AB\bigcup BC\bigcup CD$ depends on the material properties of the unbounded range Ω_0, the geometry of the boundary $AB\bigcup BC\bigcup CD$ and on the seismic load presented by the vector $\mathbf{\Phi}_1$.

To ensure the compatibility and equilibrium of tractions along interface $AB\bigcup BC\bigcup CD$, it needs the BIEM model to be converted into a macro finite element of nodal based force. In FEM, the nodal force vector due to traction t_{Γ_e} along a surface Γ_e of the $e-th$ boundary element is defined via quadratic shape functions N_i,i=1,2,3 as:

$$\mathbf{F}^e = \int_{\Gamma_e} \mathrm{N}^{\mathrm{T}}\mathbf{t}_{\Gamma_e} d\Gamma. \tag{10.27}$$

At the same time the traction $\mathbf{t}_{\Gamma_e}$ along the boundary element can be expressed by the usage of quadratic shape functions and the nodal tractions in the following form $\mathbf{t}_{\Gamma_e} = N_j t^j_{\Gamma_e}$. Having this in mind Eq. 10.27 can be rewritten as:

$$\mathbf{F}^e = \int_{\Gamma_e} \mathrm{N}^{\mathrm{T}}\mathrm{N} d\Gamma\ \mathbf{t}^e = \mathbf{\Lambda}^e\ \mathbf{t}^e \tag{10.28}$$

where the matrix $\mathbf{\Lambda}^e = \int_{\Gamma_e} \mathbf{N}^{\mathrm{T}}\mathbf{N} d\Gamma$ can be computed for each single e-th boundary element. Application of expression, Eq. 10.28, for all boundary elements along the contact interface boundary $AB\bigcup BC\bigcup CD$, leads to the global matrix $\mathbf{M} = \sum_e \Lambda^e$ and for the nodal force vector $\mathbf{F}(x \in \Gamma_{\mathrm{int}}, \omega)$ along the interface $AB\bigcup BC\bigcup CD$. Or, one can read:

$$\mathbf{F}\,(x \in \mathbf{\Gamma}_{\mathrm{int}}, \omega) = \mathbf{M}\mathbf{t}_1, \tag{10.29}$$

where the nodal traction vector along $AB\bigcup BC\bigcup CD$, is given by expression, Eq. 10.26.

After inserting of Eq. 10.29 into Eq. 10.23, the following expression is obtained:

$$\mathbf{G}_f\mathbf{F} - \mathbf{Hu} = \Phi;\quad \mathbf{G}_f = \mathbf{GM}^{-1} \Rightarrow \mathbf{F} = \underbrace{\mathbf{G}_f^{-1}\mathbf{Hu}}_{\mathbf{K}} + \underbrace{\mathbf{G}_f^{-1}\Phi}_{\mathbf{R}} \tag{10.30}$$

Finally, the nodal force vector $\mathbf{F}$ along the interface $AB\bigcup BC\bigcup CD$ is expressed via the stiffness matrix $\mathbf{K} = \mathrm{G}_f^{-1}\mathrm{H}$ Kand the free term vector $\mathbf{R} = \mathrm{G}_f^{-1}\mathbf{\Phi}$. Due to the reversal of sign of the traction continuity boundary condition, the following stiffness and force matrices $\mathbf{K} = \mathrm{G}_f^{-1}\mathrm{H}$ and $\mathbf{R} = \mathrm{G}_f^{-1}\mathbf{\Phi}$ are super-positioned into the global stiffness matrix in the FEM model.

10.3.3 Coupling the BIEM and FEM Models According to the Nodal Compatibility and Equilibrium Conditions Along the Interface Boundary AB⋃BC⋃CD

The method used to couple the BEM into the FEM environment is based on the substructure generation procedure in which a substructure of the BEM domain Ω_0 is created and attached to the FEM model of the domain Ω_1 via prescribed common nodes where compatibility and equilibrium conditions are satisfied. Substructuring is a common method in FEM simulation where a portion of the degree-of-freedom from the global equation is pre-analyzed and later condensed into a few representative one, thus, reducing the size of the main calculation.

As described above, the computation of the BEM step will result in complex-valued relation (10.32) with complex-valued dynamic stiffness matrix $\mathbf{K}_{BEM}$, displacement $\mathbf{u}$ and free-term vector $\mathbf{R}_{BEM}$ presenting the seismic load:

$$\mathbf{K}_{BEM}\mathbf{u} = -\mathbf{R}_{BEM} = \mathbf{F}_{BEM};\ \mathbf{K}_{BEM} = \mathbf{K}_{BEM_{\mathrm{Re}}} + i\mathbf{K}_{BEM_{IM}};\ \mathbf{u} = \mathbf{u}_{\mathrm{Re}} + i\mathbf{u}_{\mathrm{Im}} \tag{10.31}$$

where the notations *Re* and *Im* denote the real and imaginary parts of the corresponding vectors and matrices. In Eq. 10.31 there is no mass matrix since the effect of density ρ is represented via the fundamental solution and its derivatives.

The elastodynamic matrix equation in the frame of FEM is presented by Eq. 10.32

$$\left[-\omega^2\mathbf{M}_{FEM} + i\omega\mathbf{C}_{FEM} + \left(\mathbf{K}_{FEM} + i\mathbf{K}_{FEM_s}\right)\right]\mathbf{u} = \mathbf{F}_{FEM}, \tag{10.32}$$

where $\mathbf{M}_{FEM}$,$\mathbf{C}_{FEM}$,$\mathbf{K}_{FEM}$ are the mass, damping and static stiffness matrices,$i = \sqrt{-1}$, $\mathbf{u} = \mathbf{u}_{\mathrm{Re}} + i\mathbf{u}_{\mathrm{Im}}$ and $\mathbf{F}_{FEM} = \mathbf{F}_{FEM_{\mathrm{Re}}} + i\mathbf{F}_{FEM_{\mathrm{Im}}}$ are nodal complex displacement and complex force vectors, K_{FEM_s} is the structural damping matrix.

After fitting Eqs. 10.32 and 10.31 the following expressions are obtained for the matrices of the macro-finite element (MFE): stiffness matrix $\mathbf{K}_{FEM} = \mathbf{K}_{BEM_{\mathrm{Re}}}$, viscous damping matrix $\mathbf{C}_{FEM} = \mathbf{K}_{BEM_{\mathrm{Im}}}/\omega$, structural damping matrix $\mathbf{K}_{FEM_s} = \mathbf{K}_{BEM_{\mathrm{Im}}}$, $\mathbf{M}_{FEM} = 0$, $\mathbf{F}_{FEM_{\mathrm{Re}}} = -\mathbf{R}_{BEM_{\mathrm{Re}}}$, $\mathbf{F}_{FEM_{\mathrm{Im}}} = -\mathbf{R}_{BEM_{\mathrm{Im}}}$.

10.3.4 Illustrative Example for Seismic Response of the System: Geological Soil Region Ω_0 (BED): Soil Layer (Damper) Ω_1– Rigid Foundation Ω_f

The described above hybrid BIEM-FEM is applied only for the complex system semi-infinite BEM zone Ω_0 and FEM zone consisting of soil layer Ω_1 with rigid foundation Ω_f rested on it, see Fig. 10.3.

The seismic load is presented in Fig. 10.4 via accelerograms (a_1, a_2) recorded on May 22, 2012 in the region of the town of Pernik, Bulgaria, with magnitude 5.6 and depth 9 km.

The solutions obtained via the hybrid model described above are presented in Figs. 10.5 and 10.6. In Fig. 10.5 displacements and accelerations at the middle point along the section BC of the interface boundary $AB \bigcup BC \bigcup CD$ between the BEM zone Ω_0 and the FEM zone Ω_1 are shown. In order to verify the hybrid numerical scheme the solutions for displacements and accelerations in Fig. 10.5 are compared by the corresponding ones obtained by the pure BEM and the percentage differences are no more than 7%. Figure 10.6 draws the displacements and accelerations components of the rigid foundation Ω_f. It can be seen that the seismic signal is reduced strongly after it passes the damping layer.

The solution of the above hybrid model will give the necessary input data (displacements and forces of the rigid foundation) for modeling of the dynamic behavior of the above-ground structure rested on the foundation.

10.4 Dynamics Simulation of Structures Subject to Seismic Excitations

The hybrid use of the BIEM and FEM ensure input data for simulation of the structural dynamics in modeling the vibrations due to applied seismic excitation. Normally, the structure is simulated using finite element and the major input data are either the external forces or the forced motion of the foundation. Equations (10.32

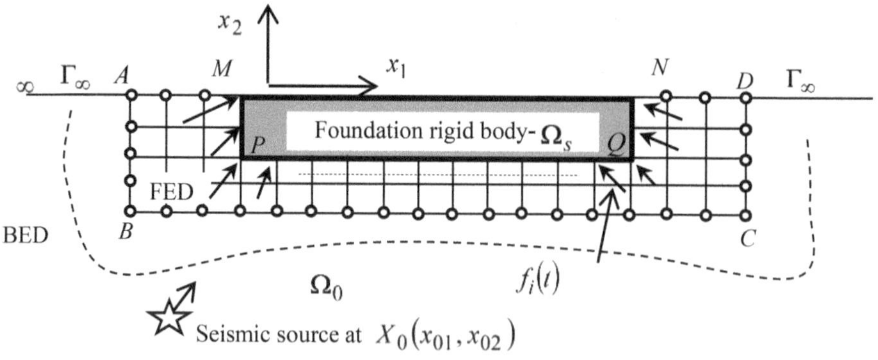

Fig. 10.3 The system: geological semi-infinite region Ω_0- soil layer Ω_1- rigid foundationΩ_f

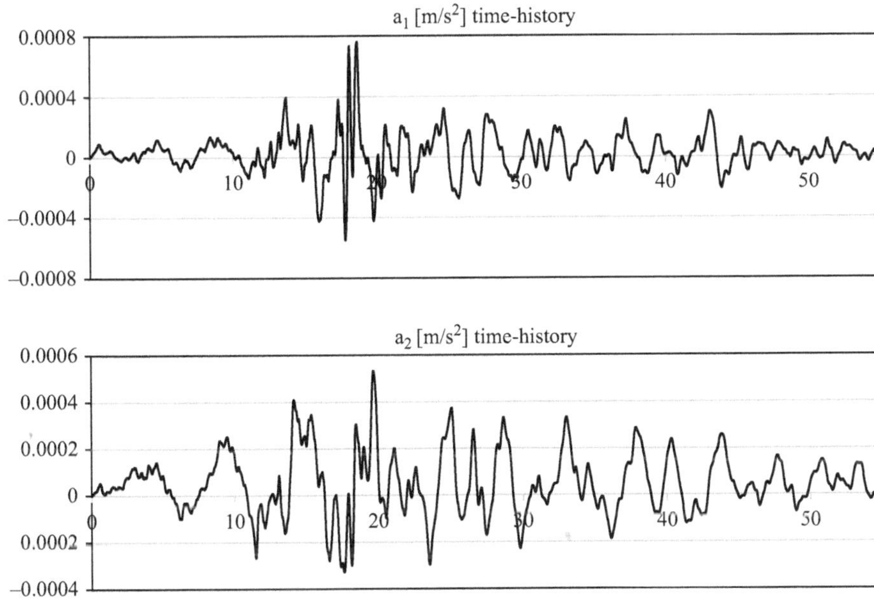

Fig. 10.4 Time histories of acceleration components in the incident seismic wave

and 10.33) define the relations of the compatible loading and displacements of the rigid foundation, taking into account additionally the mutual motion of the BED and FED domains.

Force Approach As a result of the wave propagation the forces loading the nodes of the finite element domain and of the common boundary between the layer Ω_1 and the foundation are defined. The integration procedure for ground-foundation-structure dynamics simulation consists in the following stages (Fig. 10.7):

(a) With the initial values of the displacements in the region of the seismic source, X_0 on Fig. 10.1, for which the initial displacements of the nodes in the BED and FED are zero, the displacements $\mathbf{u}$ and forces in the nodes $\mathbf{F}_{FEM}$, Eqs. 10.31 and 10.32 are computed;
(b) The matrix vector $\mathbf{F}_{FEM}$ is divided in two parts $\mathbf{F}_{FEM} = \left[\mathbf{F}_b^T \; \underline{\mathbf{F}}_{FEM}^T \right]^T$, where $\mathbf{F}_b$ is the matrix – vector of the forces loading the border of the foundation and $\underline{\mathbf{F}}_{FEM}$ are the other forces loading the nodes of the FED;
(c) With the defined already forces $\mathbf{F}_b$ and reduced to force (for the spatial case 3 × 1 sized matrix – vectors) $\mathbf{F}_{br}$ and torque $\mathbf{T}_{br}$ loading the foundation, for small increments of time, numerical integration of the dynamic equation of the flexible structure attached to the rigid foundation is implemented;
(d) With the new values of the displacements of the structure and the nodes of the FED the procedure continues from step (a) for which the new values of the forces $\mathbf{F}_{FEM}$, respectively, $\mathbf{F}_b$ are computed.

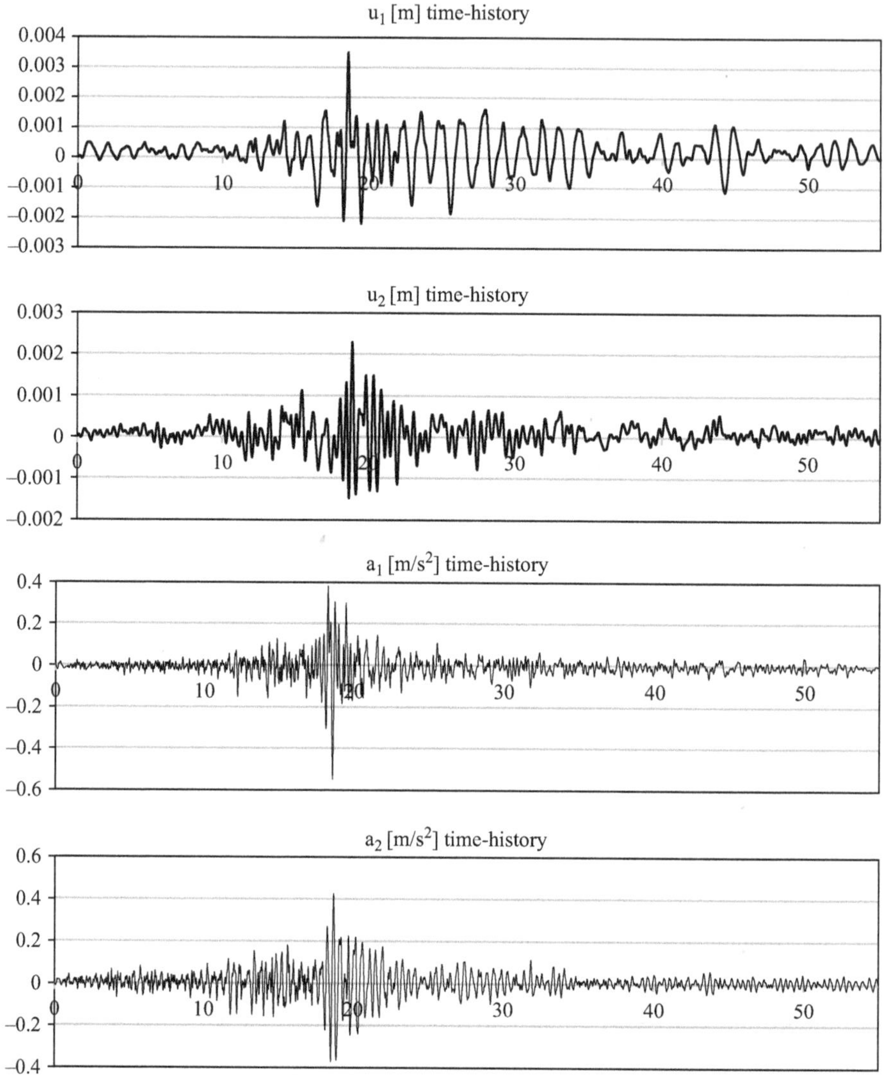

Fig. 10.5 Displacement and acceleration components at the middle point of the section BC along the interface boundary $AB\bigcup BC\bigcup CD$ between the BEM zone Ω_0 and the FEM zone Ω_1

Displacements Approach The approach is based on the foundation displacements, matrix – vector $\mathbf{U}_{br}$, as a result of the node displacements of the FED. It is assumed that these displacements define the displacements of the rigid foundation and are the initial conditions for the simulation process. In this approach to loss of precision is negligible in favor of the simplification of the integration process and decreasing the number of the degrees of freedom.

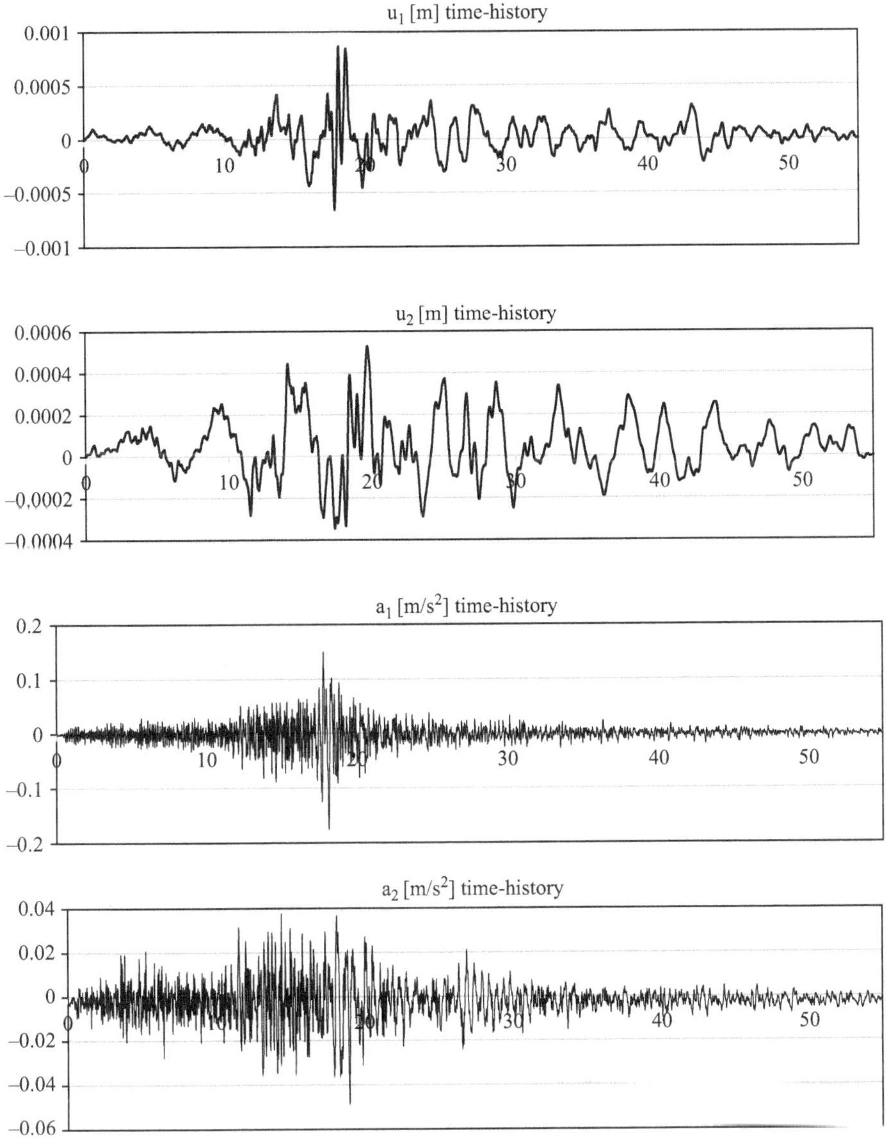

Fig. 10.6 Displacement and acceleration components of the rigid body foundation

The two procedures presented above are actually a decomposition approaches that proposes significant decreasing of amount of nodes and degrees of freedom because of solution of two different tasks, i.e.:

(a) Simulation of earthquake wave propagation and definition of the ground displacements and forces of discretized domains (Sects. 10.2 and 10.3);

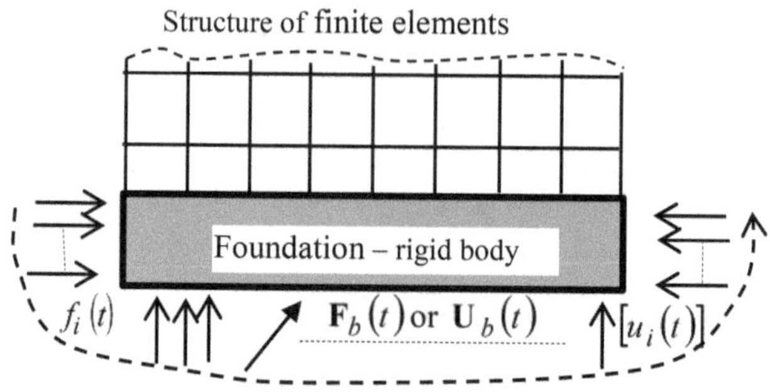

Fig. 10.7 The structure foundation with the reduced forces and displacements

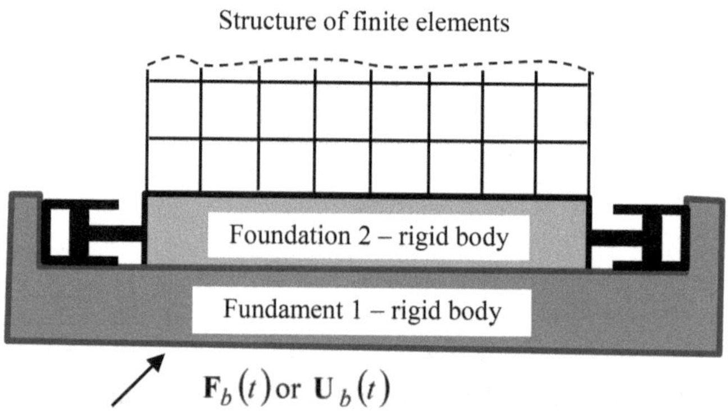

Fig. 10.8 A structural system of two foundations for active vibrations suppression

(b) Dynamics simulation of structure subject to foundation displacements $\mathbf{U}_b$ or the reduced forces and torques $\mathbf{F}_b = \left[\mathbf{F}_{br}^T \ \mathbf{T}_{br}^T \right]^T$ that load the foundation.

The procedure (b) is discussed below using as input data the results of the computations in procedure (a).

Here we discuss the possible scenarios of the procedure (b). In Fig. 10.7 the structure, the foundation and the FED are shown, where all the forces along the rigid body foundation are reduced to a spatial matrix-vector of a force which elements are the reduces forces and torques, $\mathbf{F}_b = \left[\mathbf{F}_{br}^T \ \mathbf{T}_{br}^T \right]^T$.

In Fig. 10.8 an example of a structural system of two foundations with the possibility of passive and active vibrations suppression is presented. The principle for presentation of the seismic structural excitation is similar to the case of Fig. 10.7.

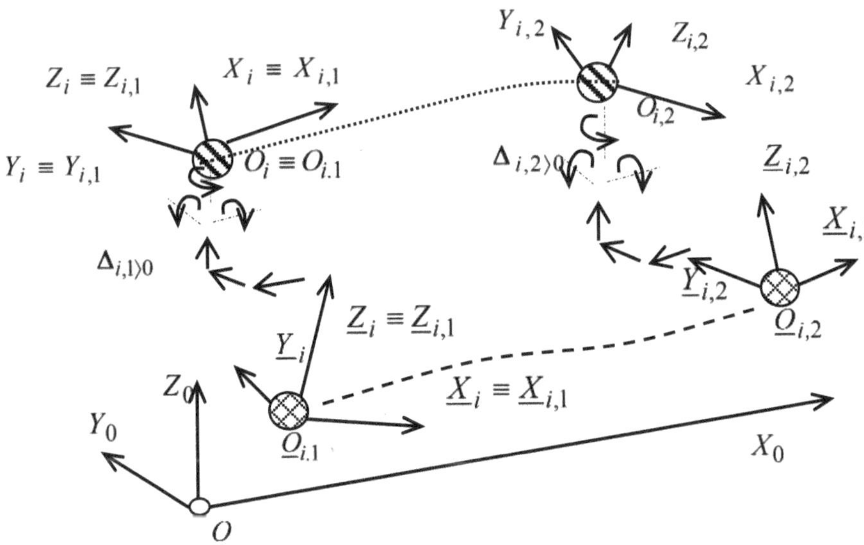

Fig. 10.9 Small global displacements with respect to the initial configuration of a flexible element

10.4.1 Dynamic Equations of a Structure of Rigid and Flexible Bodies. Finite Elements in Absolute Coordinates

The coordinate systems of the beam element and its nodes are depicted in Fig. 10.9 without the beam shape and configuration. The beam's coordinate system and that of the node i,1 coincide and their relative position does not change during the beam motion. That is why we shall regard the motion of node's coordinate systems taking into account which one is of the beam. Loading the nodes with the inertia, elastic and external forces, just like free rigid bodies they will implement, for a small increment of time, small increments of the motion parameters. Six are the minimal number of the coordinates that define the position as of a rigid body, so of a node. When a flexible element implements global motion in space overlapped by flexible displacements, even small increments of the node coordinates with respect to their initial position, do not present the real displacements of the flexible element. The solution of the problem how to use the finite element theory in case of global motion and large flexible displacements consists in:

(a) Definition of proper node reference frame coordinates for the case of small increments of time, respectively, for small increments of that coordinates;
(b) Estimation of the magnitudes of the elastic displacements;
(c) Precise formulation of the velocity dependent terms in the dynamic equations.

The solution of problem (a) consists in selection of node reference frame coordinates compatible with the coordinates used in finite element theory. The solution of problem (b) consists in separation of the flexible displacements from the global motion of the node coordinate systems. The solution of problem (c) consists in deriving generalized equations for definition of the inertia forces for rigid and flexible bodies. Regarding problem (a), in Fig. 10.9 the coordinate systems of nodes i,1 and i,2 of a non-isoparametric six dimensional beam flexible element $X_{i,k}Y_{i,k}Z_{i,k}$; i=1,2, are located in their final configuration. The initial positions of the node coordinate systems are presented by the same but underlined notations. The finite element theory (FET) claims for the small displacements to be read with respect to the system initial configuration. Here, according to the method of the FEAC [16, 20], the small flexible displacement are read with the previous configuration of the finite element. In Fig. 10.9 these displacements are presented by arrows.

Forces Loading the Nodes of Flexible Elements The forces loading the nodes are internal elastic and damping forces, external forces and, in case of the global motion, the inertia forces. For the case of the structure's foundation the reduced seismic force and torques (Sect. 10.3) are regarded as external force and torques. Regarding static structures, the internal elastic and external forces are in equilibrium. The external forces could be included in the dynamic equations using the principle of the virtual work and here we will not discuss this matter. The elastic forces are included in the dynamic equation using the stiffness matrix developed in the FET. The main subjects related to the dynamics in FET are the inertia forces. Since the nodes of the flexible elements are considered coordinate systems implementing free motion in space [20] they are kinematically independent, which means that the coordinates of a node do not depend on the coordinates of the adjacent ones. However, the static and dynamic equations cannot be solved separately for the nodes since the displacements of a node give rise of elastic forces added to the adjacent nodes. The same holds for the velocities and accelerations and the resulting inertia forces.

Inertia Forces of Flexible Elements As it was discussed above one of the major problems for application of the FET in dynamics of rigid and flexible multibody systems is definition of the node accelerations, respectively, the inertia forces. In [16] generalized Newton-Euler dynamic equations for rigid and flexible bodies discretized using the FET are derived. The inertia forces in the nodes of the flexible elements are expressed with respect to the quasi velocities and accelerations and are independent of the kind of coordinates. As general case the spatial six degree of freedom node displacements, velocities and accelerations will be regarded. The plane motion (Sect. 10.2 and 10.3) could be easily derived assuming zero displacements and their time derivative equal zero for the corresponding coordinates.

The notations for the quasi-velocities and accelerations of node i,k are $\dot{\boldsymbol{\Theta}}_{i,k} = \left[\mathrm{v}_{i,k}^{\mathrm{T}}\ \boldsymbol{\omega}_{i,k}^{\mathrm{T}}\right]^{\mathrm{T}}$, $\ddot{\boldsymbol{\Theta}}_{i,k} = \left[\mathrm{a}_{i,k}^{\mathrm{T}}\ \boldsymbol{\varepsilon}_{i,k}^{\mathrm{T}}\right]^{\mathrm{T}}$ where $\mathrm{v}_{i,k}^{\mathrm{T}}$ and $a_{i,k}^{\mathrm{T}}$, respectively, $\boldsymbol{\omega}_{i,k}^{\mathrm{T}}$ and $\boldsymbol{\varepsilon}_{i,k}^{\mathrm{T}}$, are 3×1 sized matrix – vectors of the node linear and angular velocities. For a beam flexible element i with two nodes (indices 1 and 2), and 12×12 mass matrix $\mathbf{M}$, the inertia forces $\mathbf{Fi}_{i,j}$, j=1,2, and torques $\mathbf{Ti}_{i,j}$ j=1, 2, in the nodes compile a 12×1 matrix-vector $\mathbf{F}_i = \left[\mathbf{Fi}_{i,1}^{\mathrm{T}}\ \mathbf{Ti}_{i,1}^{\mathrm{T}}\ \mathbf{Fi}_{i,2}^{\mathrm{T}}\ \mathbf{Ti}_{i,2}^{\mathrm{T}}\right]^{\mathrm{T}}$ of the beam inertia forces, i.e.:

$$\mathbf{F}_i=\mathbf{M}_i\cdot\begin{bmatrix}\ddot{\Theta}_{i,1}\\ \ddot{\Theta}_{i,2}\end{bmatrix}+\begin{bmatrix}\omega_{i,1}^{\times} & {}^{3,3}\mathbf{0} & {}^{3,3}\mathbf{0} & {}^{3,3}\mathbf{0}\\ \mathbf{v}_{i,1}^{\times} & \omega_{1}^{\times} & {}^{3,3}\mathbf{0} & {}^{3,3}\mathbf{0}\\ {}^{3,3}\mathbf{0} & {}^{3,3}\mathbf{0} & \omega_{i,2}^{\times} & {}^{3,3}\mathbf{0}\\ {}^{3,3}\mathbf{0} & {}^{3,3}\mathbf{0} & \mathbf{v}_{i,2}^{\times} & \omega_{i,2}^{\times}\end{bmatrix}\cdot\mathbf{M}_i\cdot\begin{bmatrix}\dot{\Theta}_{i,1}\\ \dot{\Theta}_{i,2}\end{bmatrix}-\mathbf{M}_i\cdot\begin{bmatrix}\omega_{i,1}^{\times}\cdot\mathbf{v}_{i,1}\\ {}^{3,1}\mathbf{0}\\ \omega_{i,2}^{\times}\cdot\mathbf{v}_{i,2}\\ {}^{3,1}\mathbf{0}\end{bmatrix}\tag{10.33}$$

where the notation "×" means skew-symmetric matrix, $\mathbf{0}$ is zero matrix and the upper left index shows the size, i.e., 3×3 is square matrix, 3×1 is matrix – vector. For elements with more than two nodes, for examples plates and shells, their elements are to be arranged in a similar order.

The general Newton – Euler dynamic equations are applicable as for rigid, so for flexible systems and provide the basis for effective general recursive algorithms and program systems for dynamics analysis.

10.5 Dynamics of Structures Subject to Seismic Excitations

The FEAC is successfully applied in this section for modelling of dynamic behavior of above-ground structure subjected to seismic excitations.

Two methods are used for simulation of seismic excitation, ground–structure force interaction method and the foundation acceleration method. The first one uses data via the simulation of the ground – structure interaction and the output data so obtained, the forces loading the foundation, are input data for the dynamics analysis of the structure. This approach is the easiest for dynamics simulation of the structures and is fully compatible with the methods discussed above.

The acceleration based approach uses statistical data from the accelerograms for specific regions and the motion characteristics of the foundation are input data for the dynamics simulation process. Using the principle of the virtual work the forces are reduced to the system coordinates to derive the dynamic equations

$$\mathbf{M}\cdot\ddot{\mathbf{u}}+\mathbf{B}\left(\mathbf{u},\dot{\mathbf{u}}\right)=\mathbf{S}\tag{10.34}$$

where $\mathbf{S}$ is $n\times 1$ matrix-vector of the generalized forces, $\mathbf{B}\left(\mathbf{q},\dot{\mathbf{q}}\right)$ is velocity depend term. The matrix – vector $\mathbf{u}$ of the coordinates includes the motion coordinates of the structure nodes and the coordinates of the rigid body foundation. In the generalized forces, matrix – vector $\mathbf{S}$ the reduced forces $\mathbf{F}_{br}$ and torques $\mathbf{T}_{br}$ loading the foundation are included as external forces. So, Eq. 10.34 is to be solved numerically with respect to unknown coordinates to simulate the structure behavior during the seismic excitation.

More difficult is the problem of solution of the differential equations, Eq. 10.34, if the displacements and their time derivatives of the foundation are input data for the simulation process. The foundation's characteristics of the response motion are presented in Sect. 10.3.3, Fig. 10.6. Since the tremendous energy of the earthquakes, the shaking effect is actually forced motion for the structure foundation. The structure vibrations cannot cause any influence on the wave propagation as in the surroundings, so in a global space. That is why, the foundation acceleration and the integrated velocities and displacements are reonomic constraints for the dynamic equations, Eq. 10.34.

An approach for solution of the dynamic equations subjected to reonomic constraints and application of FEAC [20] is presented below.

Earthquake Forced Displacements of the Structure's Foundation Using the input data for the foundation's displacement $\ddot{\mathbf{u}}_b(t)$, i.e.:

$$\ddot{\mathbf{u}}_b = \ddot{\mathbf{u}}_b(t), \tag{10.35}$$

where $\ddot{\mathbf{u}}_b = \begin{bmatrix} a_{bx} & a_{by} & a_{bz} & \varepsilon_{bx} & \varepsilon_{by} & \varepsilon_{bz} \end{bmatrix}^{\mathrm{T}}$ is the matrix – vector of the foundation linear and angular accelerations.

For a multibody system with n degrees of freedom (which motion is described by Eq. 10.34), subjected to m reonomic constraints (Eq. 10.35), the dynamic equations are presented as follows:

$$\begin{aligned} &\mathbf{M} \cdot \begin{bmatrix} \ddot{u}_1 & \cdots & \ddot{u}_{k+1} & \cdots & \ddot{u}_{k+m} & \cdots & \ddot{u}_n \end{bmatrix}^{\mathrm{T}} + \mathbf{B}(\mathbf{u}, \dot{\mathbf{u}}) = \\ &= \begin{bmatrix} S_1 & \cdots & S_{k+1} & \cdots & S_{k+m} & \cdots & S_n \end{bmatrix}^{\mathrm{T}}, \end{aligned} \tag{10.36}$$

In Eq. 10.36 the coordinates $\ddot{u}_i = \ddot{u}_i(t), i = k+1, k+2, .., k+m$ are known, while the generalized forces $S_i{=}S_i(t), i{=}k{+}1, k{+}2..., k{+}m$ are unknown. The dynamic equations, Eq. 10.36, are transformed as follows:

$$\begin{aligned} &\underline{\mathbf{M}} \cdot \begin{bmatrix} \ddot{q}_1 & \cdots & S_{k+1} & \cdots & S_{k+m} & \cdots & \ddot{q}_n \end{bmatrix}^{\mathrm{T}} + \\ &+ \mathbf{B}(\mathbf{q}, \dot{\mathbf{q}}) = -\underline{\underline{\mathbf{M}}} \cdot \begin{bmatrix} \ddot{u}_{k+1} & \cdots & \ddot{u}_{k+m} \end{bmatrix}^{\mathrm{T}} + \underline{\mathbf{S}}, \end{aligned} \tag{10.37}$$

where the matrices $\underline{\mathbf{S}}$, $\underline{\underline{\mathbf{M}}}$, $\underline{\mathbf{M}}$ are compiled according to the indices of the coordinates. The solution of Eq. 10.37 means that the direct and inverse dynamic problem is solved. For foundation – structure system it means that the reactions in the foundation and the displacements of the structure's nodes are computed.

An example of dynamics simulation of the system foundation – structure subjected to reonomic constraints (forced motion) will be presented since this is the general more difficult for application case. The structure is with significant flexibility. The aim is the possibility of the algorithm to simulate large flexible displacements and geometrical nonlinearities to be demonstrated.

10.6 Example

An example of space flexible structure compiled of rigidly connected beams that build many closed contours is presented in Fig. 10.10. The flexible beams are consequently numbered, as well as the nodes are also pointed out in ellipses. The structure is mounted on a foundation imposed on external excitation–one dimensional translational motion q_{b1} in the horizontal plane and one dimensional rotational excitation u_{b2} along the vertical axis. Because of the external excitations the foundation implements two dimensional motion which functions of the accelerations are $\ddot{u}_{b1} = a_1 \cdot b_1 \cdot \cos(b_1 \cdot t)$ and $\ddot{u}_{b2} = a_2 \cdot b_2 \cdot \cos(b_2 \cdot t)$. The structure will be simulated and analyzed for different values of the coefficients a_1, b_1, a_2 and $b_2$using the method of FEAC. The size and mass parameters of the structure are (All

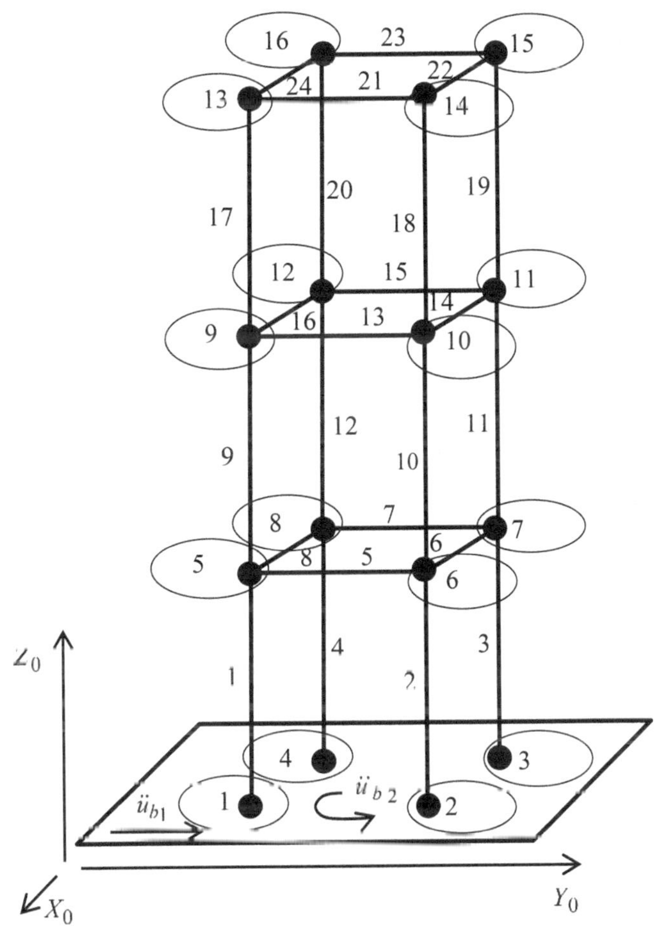

Fig. 10.10 Three story flexible space beam structure. Elements and nodes

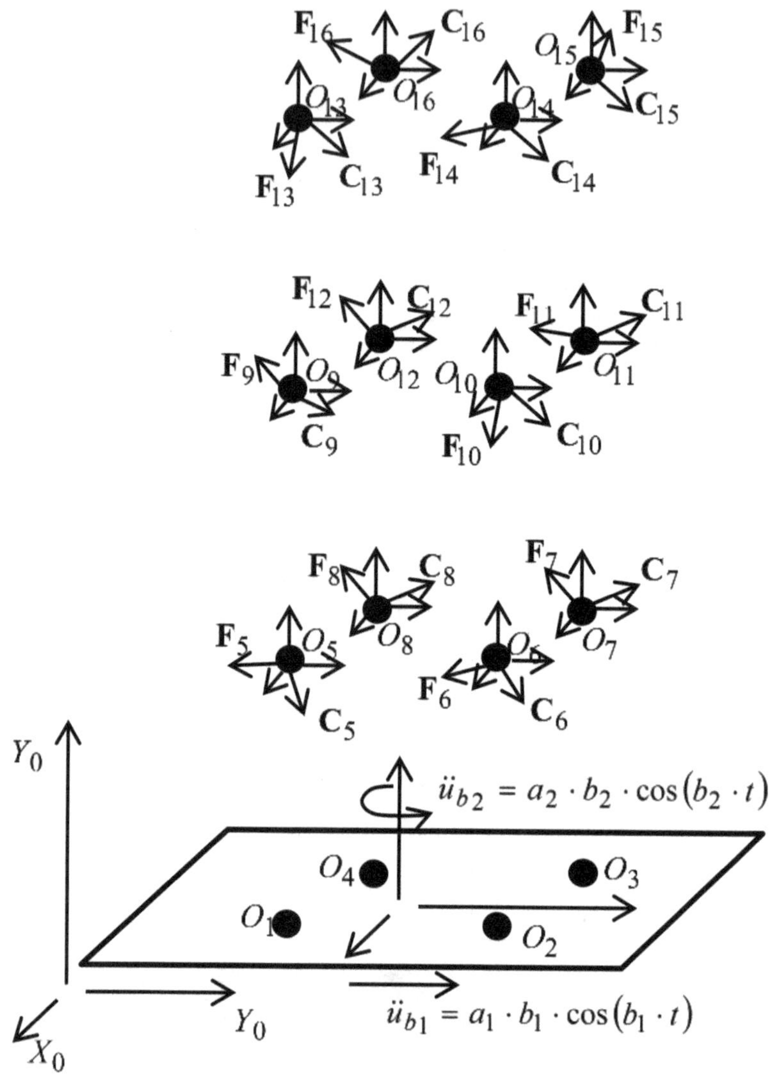

Fig. 10.11 The dynamic model of the structure as free moving coordinate systems

measures are in SI UNITS): modulus of the elasticity $E = 7\ 10^{10}$; shear modulus $G = 2.8\ 10^{10}$; length of the vertical elements $L = 10.$, of the horizontal $L = 8.$; density $\rho = 3000$; cross section area $S = 4\ 10^{-4}$; cross-section moments of inertia $I_x = I_z = \frac{1}{2} I_c = 2.10^{-7}$, foundation's mass 6000. In Fig. 10.11 the nodes of the beams are shown as moving coordinate systems which origins are pointed out as O_i,=1,2,...,16, while their axes are parallel to the absolute reference which allows the geometrical nonlinearity in dynamic behavior of the engineering structure to be

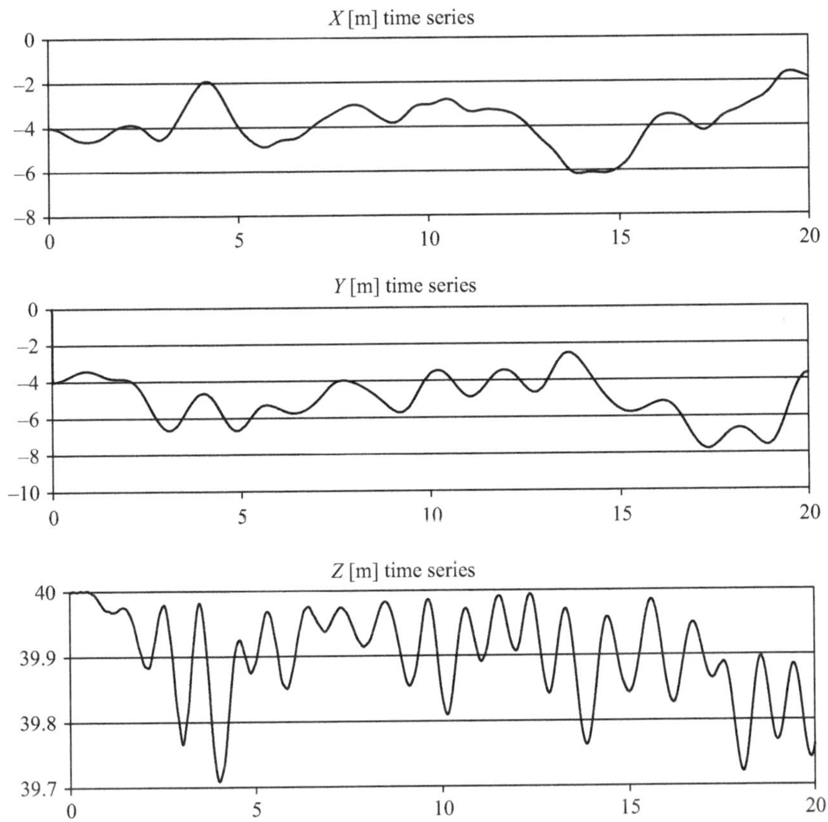

Fig. 10.12 The time histories of coordinates of node 16 for a_1=0.2 *G*; $b_1=\pi$; *a*2=0.04 *G*; b_2=1.3 π

taken into account. The nodes attached to the foundation are assumed connected to a rigid body. The nodes are loaded by elastic forces and torques caused because of the relative flexible displacements of the nodes. The nodes are also loaded by Inertia forces and torques F_i,1,2,…,16 exaggerated because of forced motion of the foundation and the resulting space motion of the whole structure. The nodes 1–4 are connected to the foundation and are considered part of the foundation as rigid bodies, i.e.: their motion characteristics coincide with that of the foundation.

The time histories of the coordinates of node 16 are depicted in Fig. 10.12. This solution is obtained for the following values of the coefficients a_1=0.2 *G*; $b_1=\pi$; a_2=0.04 *G*; b_2=1.5 π (*G* is the earth acceleration) and it satisfies the assumption that the structure's displacements do not cause building's collapse for 20 s operational time.

The time histories of the coordinates of node 16 are shown in Fig. 10.13. This solution is for the following values of the coefficients a_1=*G*; $b_1=\pi$; a_2=0.03 *G*;

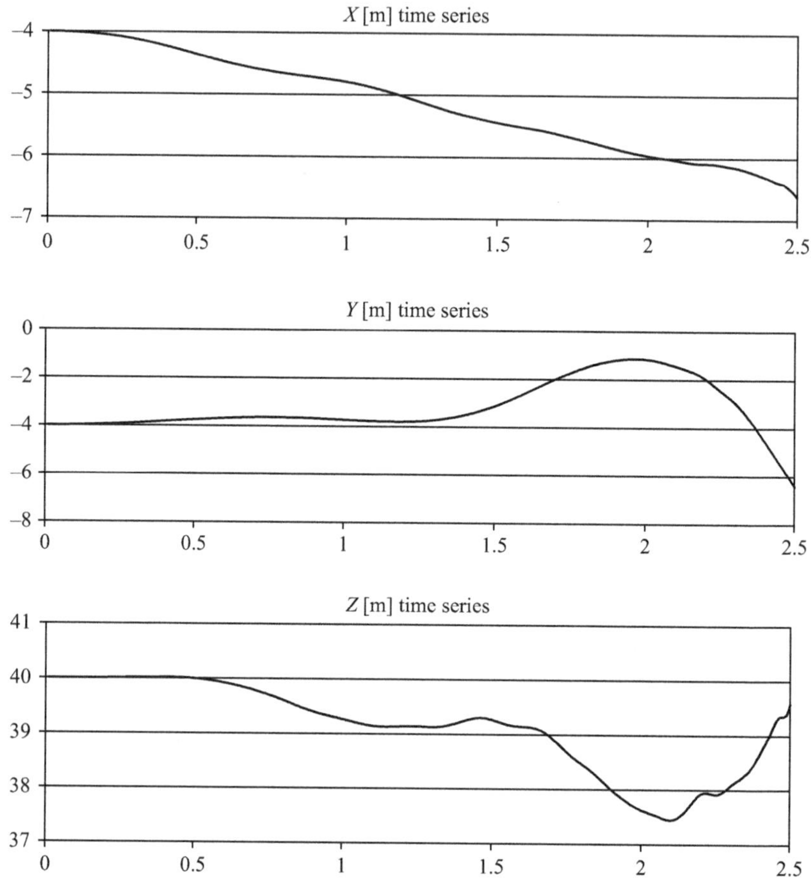

Fig. 10.13 The time histories of coordinates of node 16 for $a_1=G$; $b_1=\pi$; $a_2=0.03\ G$; $b_2=1.5\ \pi$

$b_2=1.5\ \pi$ assuming that the collapse of the structure could be observed for the third second of the operational time.

In Fig. 10.14 the configurations of the structure for three instants of time in case of the displacements for the first case (Fig. 10.12) are depicted. To be visible the displacements are scaled.

10.7 Conclusions

An efficient hybrid hi-performance methodology and software that model the dynamic response of structures during earthquake accounting for the main characteristics and mechanical properties of the soil and of the seismic source, plus the specific structural peculiarities and mechanical behavior of the building/underground structure construction is developed.

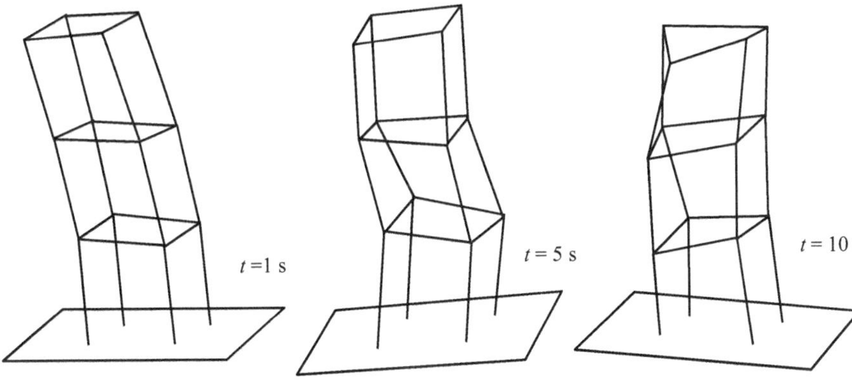

Fig. 10.14 Configurations of the structure for three instants of time in case of Fig. 10.11

The dynamic model takes into account the specific geophysical characteristics and location of the seismic source till a local region containing the engineering structure itself. A BIEM model of the semi-infinite part of the geological domain is developed, while the finite geological profile with the engineering structure rested on foundation is described via FEM. The dynamic of the above ground structure is modeled by FEAC,

It is proposed an efficient procedure of decomposition of the dynamic model regarding the whole path from the seismic source radiating transient waves to the engineering structure, followed by a subsequent analysis of the foundation's and the structure's design. Compulsive force and acceleration based approaches applied for simulation of the foundation – structure dynamics are regarded.

The approach applied proposes simulation of the seismic wave propagation from the seismic source trough a specific geological area to the structural foundation and, as a consequence, simulation of large displacements and nonlinear effects of the structure.

Examples of dynamics analysis of a structure imposed on seismic displacements are presented. Note that the developed hybrid methodology is applicable also for other type of foundation such as raft foundation, pile supported foundation, etc.

Acknowledgement The second author acknowledges the support of the Bilateral Bulgarian - Greek Project, based Personnel Program between BAS and AUTH.

References

1. Chapman, C.H.: A new method for computing synthetic seismograms. Geophys. J. R. Astron. Soc. **54**, 481–518 (1978)
2. Panza, G.F., Romanelli, F., Vaccari, F.: Seismic Wave Propagation in Laterally Heterogeneous An elastic Media: Theory and Applications to Seismic Zonation. Adv. Geophys. **43**, 1–95 (2000)

3. Fuchs, K., Müller, G.: Computation of synthetic seismograms with the reflectivity method and comparison with observations. Geophys. J. R. Astron. Soc. **23**, 417–433 (1971)
4. Kind, R.: The reflectivity method for a buried source. J. Geophys. Res. **44**, 603–612 (1978)
5. Wuttke F.: Beitrag zur Standortidentifzierung mit Oberfachenwellen. PhD thesis, Bauhaus Universitat, Weimar (2005)
6. Moczo, P.: Finite-difference technique for SH waves in 2-D media using irregular grids: application to the seismic response problem. Geophys. J. Intell. **99**, 321–329 (1989)
7. Gatmiri, B., Arson, C., Nguyen, K.V.: Seismic site effects by an optimized 2D BE/FE method. I. Theory, numerical optimization and application to topographical irregularities. Soil Dyn. Earthq. Eng. **28**(8), 632–645 (2008)
8. Gatmiri, B., Arson, C.: Seismic site effects by an optimized 2D BE/FE method. II. Quantification of site effects in two-dimensional sedimentary valleys. Soil Dyn. Earthq. Eng. **28**(8), 646–661 (2008)
9. Bouchon, M., Sanchez-Sesma, F.J.: Boundary integral equations and boundary elements methods in elastodynamics. Adv. Geophys. **48**, 157–189 (2007)
10. Dineva, P., Borejko, P., Hadjikov, L., Zigler, F.: Transient elastic waves in a half-space: comparison of the DBIE – method with the method of generalized ray. Acta Mech. **115**, 203–211 (1996)
11. Manolis, G., Dineva, P., Rangelov, Ts., Wuttke, Fr: Seismic Wave Propagation in Non-Homogeneous Elastic Media by Boundary Elements, Series: Solid Mechanics and Its Applications, Volume 240, 294 pages. Springer, ISBN 978-3-319-45205-0 (2017)
12. Manolis, G.D., Beskos, D.E.: Boundary Element Methods in Elastodynamics. Unwin and Allen, London (1987)
13. Schanz, M., Steinbach, O.: Boundary Element Analysis. Mathematical Aspects and Applications. Lectures Notes in Applied and Computational Mechanics, 29, Springer (2007). https://doi.org/10.1007/978-3-540-47533-0
14. Vasilev, G., Parvanova, S., Dineva, P., Wuttke, F.: Soil-structure interaction using BEM-FEM coupling through ANSYS software package. Soil Dyn. Earthq. Eng. **70**, 104–117 (2015)
15. Basnet, M.B., Aji, H.D.B., Wuttke, F., Dineva, P.: Wave propagation through poroelastic soil with underground structures via hybrid BEM-FEM. Z. Angew. Math. Mech. **2018**, 1–22 (2018). https://doi.org/10.1002/zamm.201800023
16. Zahariev, E.: Numerical Multibody System Dynamics, Rigid and Flexible Systems, Lambert Academic Publishing GmbH & Co. KG (2012)
17. Shabana, A.A.: An absolute nodal coordinate formulation for the large rotation and deformation analysis of flexible bodies, Technical Report MBS96-1-UIC. Dept of Mechanical Engineering, Univ of Illinois, Chicago (1996)
18. Shabana, A., Christensen, A.P.: Three-dimensional absolute nodal co-ordinate formulation: plate problem. Int. J. Numer. Methods Eng. **40**(15), 2775–2790 (1997)
19. Gerstmayr, J., Sugiyama, H., Mikkola, A.: Review on the absolute nodal coordinate formulation for large deformation analysis of multibody systems. ASME J. Comput. Nonlinear Dyn. **8**(3), 031016 (12p), (2013)
20. Zahariev, E.V.: Generalized Finite Element Approach to Dynamics Modeling of Rigid and Flexible Systems, Mechanics Based Design of Structures and Machines. Taylor & Francis Group,. 34,. 1,. 81–109 (2006)
21. Zahariev, E.: Earthquake dynamic response of large flexible multibody systems. Mech. Sci. **4**, 131–137 (2013)
22. Santana, A., Aznarez, J., Padron, L., Maeso, O.: A BEM-FEM model for the gynamic analysis of building structure founded on elastic or poroelastic soils. Bull. Earthq. Eng. **14**(1), 115–138 (2015)
23. Dominguez, J.: Boundary elements in dynamics. Computational Mechanics Publications, Southampton Boston. Co-published with Elsevier Applied Science, London New York (1993)

Index

E. Zahariev, J. Cuadrado (eds.), *IUTAM Symposium on Intelligent Multibody Systems – Dynamics, Control, Simulation*, IUTAM Bookseries 33,
https://doi.org/10.1007/978-3-030-00527-6

FSC
www.fsc.org
MIX
Papier | Fördert gute Waldnutzung
FSC® C083411

Kolibri 360 GmbH
Ferdinand-Jühlke-Straße 7
99095 Erfurt, Deutschland
produktsicherheit@kolibri360.de